Transactional Machine Learning with Data Streams and AutoML

Build Frictionless and Elastic Machine Learning Solutions with Apache Kafka in the Cloud Using Python

Sebastian Maurice

Apress®

Transactional Machine Learning with Data Streams and AutoML: Build Frictionless and Elastic Machine Learning Solutions with Apache Kafka in the Cloud Using Python

Sebastian Maurice
Toronto, ON, Canada

ISBN-13 (pbk): 978-1-4842-7022-6   ISBN-13 (electronic): 978-1-4842-7023-3
https://doi.org/10.1007/978-1-4842-7023-3

Managing Director, Apress Media LLC: Welmoed Spahr
Acquisitions Editor: Celestin Suresh John
Development Editor: Matthew Moodie
Coordinating Editor: Divya Modi

Cover designed by eStudioCalamar

Cover image designed by Pixabay

Distributed to the book trade worldwide by Springer Science+Business Media New York, 1 New York Plaza, New York, NY 10004. Phone 1-800-SPRINGER, fax (201) 348-4505, e-mail orders-ny@springer-sbm.com, or visit www.springeronline.com. Apress Media, LLC is a California LLC and the sole member (owner) is Springer Science + Business Media Finance Inc (SSBM Finance Inc). SSBM Finance Inc is a **Delaware** corporation.

For information on translations, please e-mail booktranslations@springernature.com; for reprint, paperback, or audio rights, please e-mail bookpermissions@springernature.com.

Apress titles may be purchased in bulk for academic, corporate, or promotional use. eBook versions and licenses are also available for most titles. For more information, reference our Print and eBook Bulk Sales web page at http://www.apress.com/bulk-sales.

Any source code or other supplementary material referenced by the author in this book is available to readers on GitHub via the book's product page, located at www.apress.com/978-1-4842-7022-6. For more detailed information, please visit http://www.apress.com/source-code.

Printed on acid-free paper

*For my daughter Matea, my wife Ellen, and
my mom and dad, Frank and Mary.*

Table of Contents

About the Author

Sebastian Maurice is the founder and CTO of OTICS Advanced Analytics Inc. and has over 25 years of experience in AI and machine learning. Previously, Sebastian served as Associate Director within Gartner Consulting, focusing on artificial intelligence and machine learning. He was instrumental in developing and growing Gartner's AI consulting business. He has led global teams to solve critical business problems with machine learning in oil and gas, retail, utilities, manufacturing, finance, and insurance. Dr. Maurice also brings deep experience in oil and gas (upstream) and was one of the first in Canada to apply machine learning to oil production optimization, which resulted in a Canadian patent: #2864265.

Sebastian is also a published author with seven publications in international peer-reviewed journals and books. One of his publications (*International Journal of Engineering Education*, 2004) was cited as landmark work in the area of online testing technology. He also developed the world's first Apache Kafka connector for transactional machine learning: MAADS-VIPER.

Dr. Maurice received his PhD in electrical and computer engineering from the University of Calgary and has a master's in electrical engineering and a master's in agricultural economics, with a bachelor's in pure mathematics and a bachelor's (hon) in economics.

Dr. Maurice also teaches a course on data science and actively helps to develop AI course content at the University of Toronto. He is also active in the AI community and an avid blogger and speaker. He also sits on the AI advisory board at McMaster University.

About the Technical Reviewer

 **Tim Raiswell** is a principal in advanced analytics with Loftus Labs, a consulting firm that specializes in agribusiness data science. His current areas of research include machine learning for business decision support and the importance of culture and attitude in the adoption of analytic decision-making. Tim lives in the state of Maryland with his family.

Acknowledgments

I would like to thank my wife Ellen and daughter Matea for their support in the writing of this book. Their ongoing support was invaluable.

I would like to thank Michael Scappaticci for his editing and review of the chapters and discussions on some of the concepts presented in this book. I would also like to thank Phoenix Unnayan Majumder for his inputs, especially on Apache Kafka. I would also like to thank Tim Raiswell; his insights and thoughts were important in shaping some of the ideas in this book.

Lastly, I want to thank my parents, Frank and Mary, for their continued support and encouragements in pursuing all of my dreams and goals in life.

Introduction

Fast data requires fast machine learning for fast decision-making. Understand how to apply auto machine learning to data streams with Apache Kafka in the cloud using Python, and create transactional machine learning (TML) solutions that are *frictionless* (require minimal to no human intervention) and *elastic* (machine learning solutions that can scale up or down by controlling the number of data streams, algorithms, and users of the insights). This book will strengthen your knowledge of the inner workings of TML solutions using data streams with auto machine learning integrated with Apache Kafka that are scalable. You will be at the forefront of an exciting area of machine learning that is focused on speed of data and algorithm creation, scale, and automation that will drive business value in almost every industry.

By the end of the book, you will have a solid understanding of the technical and business aspects of TML. You will know how to build TML solutions with all the necessary details, freely available software for download, all at your fingertips. You will be at the technical and business forefronts, in the knowledge economy where data creation speeds are increasing, requiring fast machine learning solutions that are frictionless and elastic, for fast decision-making that can create enormous business value on a large scale!

CHAPTER 1

Introduction: Big Data, Auto Machine Learning, and Data Streams

Data streams are a class of data that is continuously updated and captured and grows in volume and is largely unbounded [Aggarwal, 2007; Wrench et al., 2016]. Consider how our everyday lives contribute to data streams. Every time we purchase something with a credit card, the purchasing event information about your name, purchase amount, product purchased, time and date purchased, location where it was purchased, quantity, product code, and so on are all captured in real time and stored in a data storage platform capable of storing large amounts of data. Browsing the Web also results in enormous amounts of data flowing through IP networks that are being captured by your Internet service providers (ISPs). Even the cars we drive are becoming more connected to the Internet. The car manufacturers are capturing and storing all of the telemetry and GPS data.

Data continues to seep into all facets of our lives. Everyday items that we use today such as refrigerators, cars, washing machines, TVs, and so on create massive amounts of data each day. By some estimates, we create 2.5 quintillion bytes of data each day. And, most of the world's data was created in just the past few years. This is impressive in terms of scale and shows that data is flooding our world in ways we never imagined 10 or 15 years ago. Most of us are familiar with data that exists in database tables, flat files, and dataframes, but a new category of data that is creating new challenges for data engineers, scientists, and analysts is massive, fast-moving streams of data, driven by a digitally connected world. We are all aware of the growth of data and its value for organizations [Read et al., 2019; Read et al., 2020; Guzy and Wozniak, 2020; Lang et al., 2020], but we are still in the early stages of managing and analyzing fast-moving streams of data,

© Sebastian Maurice 2021
S. Maurice, *Transactional Machine Learning with Data Streams and AutoML*,
https://doi.org/10.1007/978-1-4842-7023-3_1

along with managing the load that comes with varying spikes of events that trigger large data flows which could also affect the performance of machine learning models. Data streams, or continuous flows of data, discussed in detail later, are being generated from multiple sources such as humans or machines, and the technology to manage these streams is growing. Effectively managing, and analyzing, streaming data is becoming a necessary capability in high-volume transaction industries like financial services, social technology, retail, media, health care, and manufacturing. Think of all the data generated each second (or faster) from Facebook, Twitter, LinkedIn, Netflix, IoT devices, financial technologies, and the like. These types of fast flowing data that accumulate quickly, and if permitted can grow in size to an unlimited amount, can be analyzed by data stream scientists using transactional machine learning (TML) in the following ways:

1. Data streams can be rolled back and joined to form a consolidated dataset in real time that can be used as a training dataset for TML. By analyzing "windows" of training datasets in real time, we avoid the need to analyze all the data at once; rather we can analyze all the data in transactions of time continuously. For example, if you want to analyze every retail transaction for credit card fraud, applying TML on streams of credit card transactions would help you to make decisions in time at the point of purchase.

2. Data streams can be named to help form a machine learning model. Specifically, by naming data streams, we can identify a dependent variable stream and independent variable streams to construct a model that can be estimated by TML.

3. Data streams can be repurposed to store information on the optimal algorithm that is chosen by TML. These algorithms can then be used for predictive analytics and optimization, which can be stored in other data streams and used by humans, or machines, in reports and dashboards for decision-making.

We will further show how TML leads to *frictionless machine learning* which can accelerate conventional machine learning approaches that operate on nontransactional data. Specifically, a conventional machine learning process requires human intervention when preparing data, formulating a mathematical model with the dependent and independent variables, estimating the model and fine-tuning the hyperparameters in the model, and finally deploying the model for real-world use. All of these processes cause friction that can add days or weeks to the machine learning process [Yao et al., 2019]. We show in this book how TML can significantly reduce this friction when dealing with data streams using AutoML.

We will also show how TML solutions are *elastic*. TML solutions are elastic because you can adjust the number of data streams and machine learning models that are created, as well as adjust the number of producers of data and consumers of insights from the machine learning models. This is important for several reasons:

- Allows organizations to quickly meet the analytic needs of a fast-changing business area

- Allows organizations to control costs for solutions that are no longer being used by deactivating TML solutions quickly

- Allows organizations to scale up or down solutions based on user demand

A core component of any machine learning process is data. Traditionally, a central concern for a CIO or CDO, before even doing machine learning, is developing a data strategy. But, the increase in the speed of data creates another layer of complexity for data management and analysis that is not easily incorporated in conventional data strategies. This book will provide ways to address this challenge and show how data streams can be incorporated into data strategies that will align to the goals and objectives of your organization. Before we discuss that, a question we need to ask is: what is data? A quick search for "data" on Google will bring up millions of hits on data. In this book, we will assume data that are *digitally* created. There are three forms of data:

1. Structured data

2. Semi-structured data

3. Unstructured data

Structured Data

Structured data are data that are neatly organized in some database. This structure enables developers or users to access data in a way that can be standardized and repeated for use in various types of technological solutions. Structured data is a common type of data because it makes accessing data easier for analysis and reporting. To impose structure on data, we have to do the following:

1. Classify it – Is it a number or text or image?

2. Size it – How big is the data? Or how big is it likely to get?

3. Name it – What name should we give it? Let's assume that all data with names are called "variables."

By classifying, sizing, and naming data, we are not only structuring the data but making it easier for others to use it and access it. This is important for analyzing and visualizing the data in reports and dashboards.

Semi-structured Data

These types of data have some structure to them, but not all of them are structured. Think of data that cannot completely fit into a tabular form but can be tagged and identified by keys and values. An example would be data that are called JSON[1] or XML.[2] These are generally accepted industry standard forms of labeling data by keys and values, but they do not fit in a standard, structured, relational database. Semi-structured is an important form of data because it does not require a database schema for storage. The storage of data can be defined at the application level in the form of JSON or XML. This makes semi-structured data very flexible to use, and exchange, between diverse applications which makes it easier to consume and visualize in reports and dashboards. TML solutions use JSON data formats.

Unstructured Data

These types of data have no structure. They cannot be easily classified in tabular form or put in a key-value format like JSON or XML. Unstructured data are probably the most abundant form of data because they can be created by almost any digital device. Think of emails, data from websites, video, images, sensor readings, and so on. There can be value

---

[1]JSON stands for JavaScript Object Notation; more information can be found here: `www.json.org/json-en.html`

[2]XML stands for Extensible Markup Language; more information can be found here: `www.w3.org/XML/`

in imposing a structure on unstructured data. For example, say you have thousands of emails and you want to structure the emails by keywords. Assigning a keyword to an email allows you to classify all emails with that keyword and improve the speed of searching through emails. The common thread between all these types of data is volume. The enormous growth in these types of data is referred to as big data, discussed in the next section.

A Quick Take on Big Data

The term Big Data has been used since the 1990s to describe data that are too large to be analyzed using conventional methods. Specifically, data that are not big data can easily be curated, prepped, and analyzed on a laptop or a home computer. A common definition, and origin, of the term Big Data is likely to be attributed to John Mashey [Mashey, 1999; Lohr, 2013]. His use of the term Big Data in the context of computers was the first time that someone had documented data growth and its growing demands on computer hardware such as disk space, CPU, and infrastructure, which he referred to as "InfraStress."

Big data was then characterized as [Sagiroglu, 2013]

1) Volume – This refers to the size of data that are measured in growing terabytes, petabytes, and beyond. This volume will impact the choice of storage hardware used to store these data and the types of analysis that can be done.

2) Variety – This refers to the different types of data that can be classified as big data. The types of data will impact not only the storage choice but how data are analyzed, prepped, and curated for analysis. For example, if Big Data are textual, then to perform analysis on these data using machine learning techniques will require that data be converted to numerical form for analysis.

3) Velocity – This refers to the creation speed of data. The speed of data creation will directly impact the volume of these data. This will also impact how data are processed and how they can be analyzed for insights.

4) Veracity – This refers to the quality of data. The quality of data will impact the quality of the insights that are extracted from these data. While it is difficult to gauge data quality, statistical methods and algorithms are available to determine data quality; more on data quality later.

5) Value – The value of the insights extracted from data. The value of data should be gauged within the context of the problem or area of investigation. For example, if one is trying to answer the question of why car insurance premiums are higher for younger people than they are for older people, then using data that captures the driving patterns of different demographics could add a lot of value in answering this question: optimally pricing insurance premiums for different age groups. Choosing the *right* data to address the *right* problem can offer considerable value in many business domains.

The preceding characteristics are not a complete list, but they give us guidance in understanding, characterizing, and classifying big data. Specifically, volume, variety, and velocity of data present challenges in ensuring data quality and finding ways to analyze high-speed data with machine learning that offers quality insights for decision-making. These challenges with high-speed data are exactly the ones that are addressed and resolved by TML.

Data streams can lead to big data, but big data does not necessarily lead to data streams [Jayanthiladevi et al., 2018]. Specifically, continuous flows of data will accumulate in your storage platform leading to big data. However, big data does not need to flow continuously and can be static and disk resident. Within the context of data streams (discussed later), big data characteristics that apply to data streams are velocity, volume, veracity, and variety. The value characteristic can be further applied if performing TML: when using data streams together with auto machine learning, as we will discuss in Chapter 2.

From a TML solution and infrastructure perspective, it will be important to establish an environment where data can grow and not be limited in any way. Organizations that embrace a limitless data mentality will promote a greater emphasis on data analysis to extract insights to make better data-driven decisions [Sagiroglu, 2013]. However, making good data-driven decisions will be dependent on using data with high quality. We will discuss data quality in the next section.

Data Quality

Insights that are extracted from data are directly dependent on the quality of data. Data quality concerns do not change between conventional, static data and continuously flowing data streams. However, how quality issues are determined and identified does vary between the two types of data, and this is directly a function of the velocity of the data. For example, the higher velocity of data will give rise to faster changes in the underlying structure of the data. Detecting and improving data quality in data streams present further challenges. Given the continuous flow of data, assessing quality requires automated and real-time processing [Gudivada et al., 2017]. How can you perform data imputation or detect duplicate data in a data stream? This is still an outstanding issue but is starting to get more attention [Gudivada et al., 2017]. TML can offer some help in this area, as discussed in Chapter 2. Specifically, if trying to identify outliers, or anomalies, in the data, conventional anomaly detection mechanisms may not pick up all outliers. The issue with conventional approaches is that they do not take into account transactional data that vary with time. We will show how TML uses unsupervised learning algorithms to detect outliers, or anomalies, in fast flow data streams that vary quickly with time.

The adage of garbage in, garbage out is true. Good data, as opposed to bad data, is a critical requirement for good insights. But how does one determine whether data they have is of good quality? Using the International Organization for Standardization (ISO) definition on quality [Heravizadeh et al., 2009]: *the totality of the characteristics of an entity that bear on its ability to satisfy stated and implied needs*. These stated and implied needs will vary based on the environment and context in which these data are being used. This would further imply that data quality thresholds will vary with the environment and context. For example, it is likely that the data quality threshold needed to measure someone's risk of cancer would be higher than the quality threshold needed to measure the likelihood of people preferring Coca-Cola over Pepsi.

In fact, the dimensions of data quality will vary as well. For example, in accounting and auditing, *accuracy*, *relevancy*, and *timeliness* are three important data quality dimensions [Sidi et al., 2012]. In the area of Information Systems, *reliability*, *precision*, *relevancy*, *usability*, and *independency* are important. Table 1-1 shows a consolidated list of data quality dimensions [Sidi et al., 2012, pp. 302].

Table 1-1. *Data Quality Dimensions*

Dimension	Definition
Timeliness	The degree to which the age of the data is appropriate for the tasks to be executed. Specifically, timeliness is the delay between a change of a real-world state and the resulting modification of the information system state. This can affect its age and volatility, such that age measures how old the information is relative to when it was recorded, and volatility is a measure of the instability of the information based on the frequency of the change in the value in the entity attribute.
Currency	The degree to which the datum is up to date despite discrepancies caused by time-related changes to the current value. Currency is a description of when information was entered in sources or data storage technologies like databases and the time it was entered.
Consistency	The degree to which the data is presented in the same format and compatible with previous data.
Accuracy	The degree to which the data is correct, reliable, and certified. Specifically, accuracy measures the proximity of a data value to some other value that is considered correct.
Completeness	The degree to which the data is considered to be a representation of the real-world system, such that the data has enough breadth, depth, and scope for the tasks to be executed and contains all of the values that are supposed to be collected as theory would suggest.
Accessibility	The degree to which the data is easily available and quickly retrievable.
Duplication	The degree to which the data has duplication existing within or across systems for a particular field, record, or dataset.
Data specification	The degree to which the data can be considered as complete, well documented, with data models, business rules and metadata, and reference data.
Presentation quality	The degree to which the data can be used for visualization and supports appropriate use of visualization techniques like graphs and dashboards.
Consistent representation	The degree to which the data achieves a standard format that can be used across applications and systems.

(continued)

Table 1-1. (*continued*)

Dimension	Definition
Reputation	The degree to which the data is highly regarded in terms of its source and content.
Safety	The degree to which the data achieves reasonable and acceptable levels of risk of harm to humans, property, process, and environment.
Appropriate amount of data	The degree to which the volume of data is appropriate for the tasks to be executed.
Security	The degree to which the data is restricted appropriately to maintain its security.
Believability	The degree to which the data is regarded as true and credible.
Understandability	The degree to which the data is clear, unambiguous, and easily comprehended.
Objectivity	The degree to which the information is unbiased, unprejudiced, and impartial.
Relevancy	The degree to which the information is applicable and helpful to the tasks to be executed.
Effectiveness	The degree to which users can achieve specified goals and objectives, with accuracy and completeness in the context of the use case.
Interpretability	The degree to which the data can be interpreted in different languages, symbols, and units with clarity.
Ease of manipulation	The degree to which the data are easy to manipulate and transform.
Free of error	The degree to which the data are reliable and correct.
Ease of use and maintainability	The degree to which the data is easy to use but can also be easily accessed, updated, maintained, and managed.
Usability	The degree to which the information is clear and easy to use.
Reliability	The degree to which the information is correct and reliable enough to be used for analysis without negatively impacting the performance and quality of results.
Amount of data	The degree to which the quantity of data available is appropriate for analysis.

(*continued*)

Table 1-1. (*continued*)

Dimension	Definition
Freshness	The degree to which the family of quality factors, with each one representing some degree of freshness aspect influencing its metrics.
Value added	The degree to which the information provides advantages from its use.
Learn ability	The degree to which the information captured can be used for learning.
Data decay	The speed to which the data are negatively changing [McGilvray, 2008].
Concise	The degree to which the information is not overwhelming yet complete and to the point.
Consistency and synchronization	The degree of equivalence of the information in other data stores, applications, and systems and the processes for making data equivalent.
Data integrity fundamentals	The degree of the existence, validity, structure, content, and other basic characteristics of the data.
Navigation	The degree to which the data are easily linked.
Useful	The degree to which the data are helpful for the tasks to be executed.
Efficiency	The degree to which the data are able to meet the information needs for the tasks to be executed.
Availability	The degree to which the data are fully accessible.
Data coverage	The degree to which the data represent the entire population of data.
Transactability	The degree to which the data will produce the desired business transaction or result.
Timeliness and availability	The degree to which the data are current and available for use in the timeframe needed for analysis.

Using data mining and statistical techniques, together with finding dependencies between dimensions, allows us to determine the level of data quality [Sidi et al., 2012]. But, assessing data quality in Big Data offers some challenges such as dealing with complex factors, missing data, data duplication, and data heterogeneity all resulting from data being sourced from multiple sources [Gudivada et al., 2017] and generated by multiple types: humans and machines. Several data mining and statistical techniques are available to improve data quality such as data imputation to fill in missing data, outlier detection using machine learning algorithms like regression analysis, and duplicate data detection using natural language processing [Gudivada et al., 2017].

Finding dependencies between dimensions and then using data mining and statistical techniques to objectively measure the level of quality offers promise. We can apply a framework that shows how dimensions are related to data variables that would lead to a data quality improvement [McGilvray, 2008]. However, while it is beyond the scope of this book to delve too deeply into data quality dimensions, it is important to be aware of potential data quality issues before conducting any analysis.

The most common data quality dimensions are accuracy, currency, consistency, and completeness. When using data for analysis, at a minimum, consideration should be given to these four dimensions to gauge the level of quality of data. Figure 1-1 [Sidi et al., 2012] shows a general framework in assessing data quality based on independent and dependent variables that are being considered for analysis. With data streams, these dimensions are less amplified, putting greater emphasis on the consistency and completeness of the data for TML.

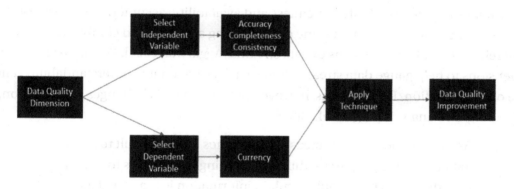

Figure 1-1. *Data Quality Framework [Sidi et al., 2012]*

As shown in Figure 1-1, the separation of the independent and dependent variables highlights the differences in the types of data quality dimensions needed for different types of variables being considered. This also shows that careful thought needs to be given when choosing the dependent and independent variables – as quality dimensions are likely to differ. As mentioned earlier, applying statistical techniques to gauge data quality on transactional, or event-driven, data will present challenges such as getting enough historical data to analyze, choosing the right subsets of data, comparing data between subsets, and applying statistics to these subsets in real time.

In this book, we will look at how TML is used to extract subsets of data in real time that can be used for machine learning, which could aid in detecting data quality issues in transactional data. Specifically, we will show, in Chapter 6, how supervised and unsupervised machine learning models can be constructed and estimated in real time to address a wide range of business problems and data quality issues, like anomaly detection.

While there are challenges in gauging data quality in data streams, the preceding dimensions lend guidance in showing us what types of characteristics should be considered when addressing data quality issues. Using TML in addressing issues of data quality in data streams can help reduce some of the challenges and improve the quality of insights we can get from data streams. The next section introduces data streams.

Data Streams

Data streams create several challenges for data analysis. First, how do we determine data quality? From Table 1-1, we know that the data quality dimensions most applicable for data streams are transactability, timeliness and availability, accuracy, completeness, freshness, and consistency. With some authors using accuracy and completeness as the most relevant quality dimensions of data streams [Aggarwal, 2007]. But there are other dimensions to help gauge data stream quality such as reliability, understandability, and ease of manipulation. Data streams, however, present several challenges for collection, processing, and analysis [Aggarwal, 2007]:

- As the volume of data increases, it becomes more difficult to process data with multiple passes, thus constraining the analysis to a single pass through, which could result in information loss as the data accumulates.

- As the data evolve over time, they exhibit *temporal locality*. This means that any data analysis using data mining algorithms must have a clear focus on incorporating the influence of time on changing data.

- Data streams are distributed in nature, meaning there will be multiple streams of data that must be accounted for and therefore considered in the analysis. This further complicates the process of using data streams in any analysis and also makes it more difficult to gauge the quality of multiple streams.

Analysis of data streams, using machine learning, is further complicated by business problems that require real-time solutions for faster decision-making such as fraud detection, product recommendations, autonomous device control, and so on. These real-time solutions will require real-time processing of data streams which could present memory, CPU, and throughput challenges for local systems [Cormode, 2005; Kollios et al., 2005]. Using cloud-based infrastructure lessens these complications because cloud infrastructure is more *elastic* in its ability to scale horizontally as data grows, as well as to scale vertically if more CPU power is needed. An approach to identifying the right business problems will be discussed in Chapter 4.

The elasticity of cloud infrastructure is important in managing, analyzing, and storing data streams. While the cost of data storage is becoming less of a concern today, doing machine learning on data streams presents other unique challenges such as constructing training datasets to deliver the analytic insights in a real-time, intuitive format for decision-making purposes. Some of these challenges are discussed in the next section.

Stream Mining

Extracting value from data streams requires the ability of a machine to gather the right data streams, combine them to form a training dataset, and then apply an algorithm to the training dataset. Given the speed of data accumulation in the streams, it is impossible to perform conventional machine learning, which requires extensive human involvement, on the combined stream of data. We refer to the degree of human involvement as **friction**: high degree of human involvement leads to a high level of friction, and a low degree of human involvement leads to a low level of friction in the machine learning process. Recent research shows a growing interest in how auto machine learning (AutoML) leads to a reduction in human involvement and accelerates the outcomes from machine learning for use by humans and machines [He et al., 2020; Yao et al., 2019; Amershi et al., 2014].

One of the main goals of TML is to reduce this friction and promote a frictionless machine learning process. While applying machine learning to data streams, as proposed in this book and demonstrated in Chapter 6, is still a new phenomenon, past work has laid the foundation for some of the concepts used in TML. Some of these foundational areas are in stream mining [Aggarwal, 2007] such as

- Data stream clustering – While clustering techniques are widely applied to conventional data, applying cluster to data streams presents challenges due to the one-pass constraints on the dataset. Some have used micro-clustering techniques to data streams, by clustering the entire dataset, but this could add strain on computing and performance as the dataset grows.

- Data stream classification – Classification techniques are widely used in machine learning modeling. Specifically, classification requires the classification of the values in the dependent variable based on certain business rules or logic used to justify the classification. By classifying the data, algorithms that are applied to the model can be used to predict the probability or likelihood of each class occurring given values of the independent variables in the model. But classifying the dependent variable is complicated with data streams because it requires past knowledge of the classification which is almost impossible to determine with real-time data, but much easier with static data. We will show in this book how unsupervised learning can be used to detect the likelihood of anomalies or fraud in the presence of temporal locality that can help you alleviate the issue of classifying the dependent variable stream.

- Frequent pattern mining [Agrawal et al., 1993] – Data streams are data that change frequently and differ from *conventional* data that are static and disk resident. Therefore, finding ways to analyze data that change frequently offers challenges and opportunities for TML that will be discussed in this book.

- Change detection in data streams – Given the nature of data streams, many times it will be important to track and analyze the changes in the trends and patterns and how they can impact the analysis. This has been identified as a challenge for analysis because the changes in data streams can also impact the change in the analysis [Agrawal, 2003; Dasu et al., 2005; Kifer et al., 2004]. For example, if you are analyzing customer data, detecting when a customer's address changes is complicated with data streams because you need to analyze past addresses for each customer and then use in-memory logic that can detect a change in the address and then perform some action. In this book, you will learn techniques that will help you with these types of use cases.

- Load shedding in data streams – Given data in data streams are produced from external sources, it is difficult to control the incoming stream rates; load shedding is a way to balance the incoming streams by degrading performance for unused streams. In this book, we show another way to manage streams by tracking who is producing to the stream and who is consuming from streams; if no one is consuming from the stream, then we show how this stream can be deactivated, thereby reducing unnecessary usage of storage, CPU, throughput, and reducing overall network load.

- Sliding window computation in data streams [Datar et al., 2002] – Most data streams are de facto time series data that change with time, which complicates historical training dataset creation. This is solved using a sliding window approach that slices the data stream in windows of small time increments to construct the training data. Computations are then performed on this data window slice that reduces computational overhead and improves performance.

- Synopsis construction in data streams – Further addressing the challenges with analyzing large volumes of data, techniques [Garofalakis et al., 2002] can be used to generate a synopsis of the entire dataset. For some problems, it may be acceptable to get an approximate solution to the problems by constructing a synopsis. In this book, we will show how synopsis can be used to construct a training dataset for machine learning that allows for deeper understanding of the data and allows you to perform predictive analytics, anomaly detection, and optimization with data streams in real time.

- Join processing in data streams – Joining streams is critical to correlating information between streams. Joining streams from different sources adds additional dimensionality to the data, making the joined data streams more effective for machine learning modeling. In this book, stream joining will play an important part in transactional machine learning because joining streams allows you to construct training datasets that will be used by TML. A challenge in stream joining is maintaining consistency in the state of the data across streams; it is however impossible to maintain a consistent state due to the fundamental nature of data streams such as temporal locality [Aggarwal, 2007].

- Indexing data streams – While important in conventional data, with data streams indexing the data presents challenges due to the increasing volume of data. Properly indexing data could improve the tracking and analysis of data. In this book, we use key and value pairs to uniquely represent each datum at the time of creation to improve the storage and retrieval of large datasets in the cloud.

- Dimensionality reduction and forecasting in data streams – Data accumulation, over time, in streams is already challenging for analysis; this is further complicated when dealing with multiple data streams, thus requiring dimensionality reduction [Sakurai et al., 2005; Yi et al., 2000]. Reducing the dimensionality of data streams will be important for effective and efficient modeling. Theory should drive the selection of variables that formulate a machine learning model; TML does not remove this requirement. It will be important to ensure unnecessary variables are not included in the machine learning model by knowing, beforehand, which streams are important and which streams are not important.

- Distributed mining of data streams – Data streams can originate from different sources or nodes and underlying processes and be stored in different locations. Dealing with this distributed nature of data streams presents challenges for storage, CPU, memory, and network throughput at each node. Managing distributed data also adds communication overhead across the network. This type of scale is not uncommon with data streams especially when organizations are large multinationals. In this book, we will show how this can be effectively managed with proper storage and retention strategies that make it easier to manage a large number of distributed data streams.

The preceding aspects of data streams can make them seem difficult to use for analysis. For example, while one data stream may be easier to handle, dealing with multiple data streams raises challenges with load management, data quality, stream joins, and so on. But, as we will show in Chapter 5, choosing the right technology and architecture can dramatically improve the way you can manage and analyze data streams. By using an application programming interface (API) and microservices-based architectures, you can scale TML solutions both horizontally and vertically, making them ideal for large-scale deployments.

In the next section, we discuss how auto machine learning can be integrated with data streams, enabling us to apply advanced machine learning algorithms to the machine learning model to find an optimal algorithm that best fits the data in real time. This optimal algorithm can then be used to make predictions and perform optimization.

Auto Machine Learning (AutoML)

Up to this point, our focus has been on data quality, data streams, and some of the important characteristics they possess; this will be important in understanding how to build TML solutions. We now discuss how we can extract value from data streams with AutoML.

AutoML, as the name indicates, is automating the process of building machine learning models that require little to no human intervention [Yao et al., 2019]. Mitchell (1997) said it best: "Ever since computers were invented, we have wondered whether they might be made to learn. If we could understand how to program them to learn – to improve automatically with experience – the impact would be dramatic."

Xin et al. (2020) provide a good summary of AutoML across a wide range of research papers. Because we are automating machine learning, we have to ask: what is machine learning? A simple way to think about machine learning is to use the human learning analogy. As humans, we learn from experiences and information given to us by our parents, teachers, friends, social groups, and others around us in different environments. The human mind processes and stores this information or data as learnings, and we use these learnings to help guide our decisions and choices throughout our lives within our environment. Similarly, a machine uses historical data that ranges over time and represents a combination of consolidated data that are named, signifying variables; this combination of variables is structured in the *right way* to form a model made up of dependent and independent variables. We emphasize the *right way* because this requires, as mentioned, some human belief or theory that guides the selection of which variables are the influencing factors (i.e., independent variables) on the dependent variable (i.e., response variable).

The rationale for forming a model with variables is to answer some question or solve some problem. If the problem is to answer the question "What will the weather temperature be tomorrow?", as humans, we can answer this question by looking out of our windows or going outside and make a guess about what the weather temperature is likely to be tomorrow. Or, we can take a more data-driven approach by gathering data from the past year and formulate a model that uses *relevant* variables to help us predict tomorrow's temperature. We use a dependent variable, the variable we are trying to explain such as weather temperature, and independent variables, variables that are *believed* to influence the dependent variable. This belief will be based on our experience on factors influencing weather temperatures, what experts say about the variables influencing temperature, theory about weather temperatures, and so on. Using this information, we can gather these variables and formulate a weather temperature (WT) model along with other independent variables that we believe will influence WT such as cloud cover (CC) and time of year (ToY).

Next, we can represent the model in an estimated form, as shown in Eq. 1 as

$$WT = a + b*CC + c*ToY \qquad\qquad (Eq.\ 1)$$

that a machine can use to determine how the independent variables, mathematically, relate to the dependent variable by estimating the parameters a, b, and c. Once the machine has estimated a, b, and c, we can then use the estimated equation to predict what the weather (WT) is likely to be tomorrow. So, in a very simple example, we created a model, using variables, from which a machine can learn: the learnings are captured in the estimated coefficients a, b, and c. But, how does it learn? Learning by machines is done using algorithms, which are set of instructions based on logic, to help the computer determine how, mathematically, the independent variables (CC and ToY) correlate with the dependent variable (WT). This correlation, or learning, is represented as estimated coefficients (a, b, c) on each of the independent variables, as shown in Eq. 2. For the sake of example, if the machine, using algorithms, has chosen a=20, b=1.2, c=1.5, then Eq. 1 becomes Eq. 2.

$$WT = a + b*CC + c*ToY \qquad\qquad (Eq.\ 1)$$

$$WT = 20 + (1.2)*CC + (1.5)*ToY \qquad\qquad (Eq.\ 2)$$

Now, if you want to know what WT will be tomorrow, you have an objective way to make that prediction by simply plugging in values for CC and ToY. If CC=2, and ToY=6 (June), then the predicted temperature is shown in Eq. 3:

$$WT = 20 + (1.2)*(2) + (1.5)*(6) = 31.4 \qquad \text{(Eq. 3)}$$

which is equal to 31.4 Celsius. The main challenge with the preceding example is ensuring the estimated values for a, b, and c are the *best* representation of the correlation to the dependent variable. A human data scientist solving the weather problem might manually iterate through 20 models or more, checking their predictions, before settling on the best one. This is expensive both in terms of time consumed and often in terms of model quality. AutoML can trial many more models almost instantly before settling on the model or models that fit the problem best. In the context of AutoML, the best can be determined by using statistical metrics that measure the accuracy of the model's predictive power when compared to actual values. This predictive power measures how close the predicted values are to out of sample or actual values; the closer the predicted values are to actual values, the *better* the algorithm and estimated coefficients, within the context of the data used. A popular metric to measure this closeness is the MAPE[3] (mean absolute percentage error). Imagine if there was a way to go through any and all algorithms, fine-tune the hyperparameters in the algorithm, compute the MAPE, and then choose *that* algorithm with the lowest MAPE as the best algorithm: this is what AutoML can help with and is an important aspect of AutoML [Xin et al., 2020]. Auto retraining and tuning is especially important with data streams because models can degrade quickly with fast data. By retraining and retuning more often with slices of data windows, the likelihood of better quality insights is higher. Furthermore, TML is technically performing machine learning on big data without the need for additional software or hardware. In Chapter 6, we will demonstrate how this is done.

AutoML is a fast-growing area of interest for both academics and organizations [Xin et al., 2020]. Especially as the growth and speed of data creation continues to accelerate and organizations become more aware of the value of data and the insights that can be extracted, they will continue to find faster ways to extract value from data. Extracting value fast from data streams is a core focus of TML. Specifically, we will show that combining data streams with AutoML, with an architecture that can scale,

---

[3]The MAPE is a ratio of the absolute difference between the predicted values and actual values in the numerator, divided by actual values.

can help to meet the demands of organizations for faster machine learning [Gudivada et al., 2017]. As data flows get faster, the need for faster insights grows, and the ways to use those insights for faster decision-making become important to maximize business value. Within a distributed network of data streams, large organizations will need to not only balance data loads but also ensure data is secure and cyberthreats are effectively managed by maintaining strong oversight and governance of both data and algorithms. In Chapter 6, we will show how governance and oversight of algorithms can be accomplished using the Algorithm and Insights Management System (AiMS) technology. The next section discusses the machine learning model building process as it relates to data streams.

Machine Learning Model Building Process

The conventional process of model building for machine learning requires humans to gather data and variables, perform analysis of data to extract important features, formulate a model with those features, estimate the model, test the model results by conferring with experts on the validity of the results, verify the correctness of the model formulation, and then deploy the model for human or machine consumption. This process will take time, effort, and money that may present limitations for many organizations and may not be suitable for data streams. We will discuss how TML can lead to a *frictionless machine learning* process that reduces the need for human touchpoints in the model building, estimation, and deployment stage. This reduction in friction aligns with data stream mining and analysis in three ways:

1. Speed – Speed to insights from data streams is complicated by temporal locality of the streamed data, which makes conventional machine learning challenging. TML accelerates time to insights from data streams with AutoML.

2. Scale – Streaming data accumulates fast resulting in big data. Handling entire streamed datasets is not practical for reason presented before, but applying machine learning to transactions of datasets in the form of sliding windows can dramatically improve the degree and depth of insights from streamed data.

3. Depth of insights – Data mining is "the analysis of (often large) observational datasets to find unsuspected relationships and to summarize the data in novel ways that are both understandable and useful to the data owner" [Hand et al., 2001]. It is also "the search for valuable information in large volumes of datasets" [Weiss and Indurkhya, 1998]. Data mining differs from machine learning which aims to *teach* the computer, using historical data, the trends, patterns, or correlations between variables that are carefully formulated in a mathematical model – learning by computers is part of artificial intelligence. Arthur Samuel, a pioneer in machine learning, defined machine learning as a "field of study that gives computers the ability to learn without being explicitly programmed" [Samuel, 1959]. As shown in our simple WT model, the machine captures these learnings in the form of numerical estimates of a, b, and c parameters or coefficients. Another difference between machine learning and data mining is the former can learn from multidimensional data variables formulated in a model, while data mining is best suited for large one-dimensional datasets.

TML (the application of auto machine learning to data streams will be defined formally in Chapter 2) is a new area of investigation that provides deeper insights from data streams. It reduces the friction that exists in conventional machine learning processes while increasing the elasticity of machine learning solutions, to accelerate, at scale, the consumption of insights for decision-making by humans or machines in real time. TML opens up exciting opportunities for academics and organizations by offering a way to learn deeper patterns and structures in data streams by formulating TML models that can be estimated in real time and used for predictive analytics and optimization, as well as for data quality improvements. The next section concludes with final remarks and a description of the chapters to follow.

Concluding Remarks

The increasing levels of data from different sources continue to grow. Almost everything we use or wear produces some sort of digital data that gets captured and consumed by humans or other machines. The advent of faster computers, larger data storage facilities, dropping costs for computing and storage, and dramatic growth of cloud-based infrastructures that use "as-a-service" business models is making advanced analysis of big data streams accessible. Infrastructures that would have been out of reach for many are now available for a few dollars a month. In fact, the growth of advanced cloud infrastructure has made it possible for anyone to perform advanced analysis like machine learning on large (and small) datasets without incurring large costs. This is leading to a democratization of machine learning and data science that is driving innovation and disrupting almost every field and industry.

There are however growing challenges that we aim to address in this book. First, very little work has been done in the area of combining data streams with AutoML to build frictionless and elastic machine learning solutions. We suspect the reason is that AutoML is a new area that has yet to be fully exploited [Xin et al., 2020], let alone with data streams. Second, big data is also a recent phenomenon, as most of the world's data has been generated in the past few years, but accelerating very quickly due to the increased demands from consumers for more real-time data. This is leading to organizations streaming more data to their customers;[4] as a result, the need to do deeper analysis using machine learning on data streams is getting more important that will enhance, and create, new products and services that improves customer experiences and reduces costs. Third, as costs for big data storage and compute continue to decline, analyzing large datasets offers more opportunities to uncover insights that can offer organizations better intelligence about who their customers are and what they like or dislike; this can be extremely valuable for targeted marketing and advertising campaigns and is exactly the business practice followed by Facebook and Google [Carter, 2013]. Intelligence is commonly unearthed using data mining techniques, but using machine learning to find deeper insights by combining other data streams is still not widely applied by organizations [Xiang et al., 2019; Jayanthi and Sumathi, 2016]; this book shows how you can apply auto machine learning to data streams that can create tremendous business value fast.

This chapter presented an introduction to the main areas of transactional machine learning – big data, data streams, and AutoML – and the unique challenges that occur when combining them such as data quality issues, lack of speed to insights, scalability and load shedding issues, and depth and quality of insights. It is widely accepted that extracting insights by analyzing data is fundamental to improving many areas of our personal and professional lives. Data mining and machine learning techniques and methods are well established. Current and future trends are all pointing to faster creation of data requiring faster ways to extract deeper insights from these data.

---

[4]Netflix is a good example of this, as well as music streaming companies like Apple and Spotify.

Chapter 2 delves deeper into TML by defining it and presenting the principles of TML. It brings all the pieces together to show why it will become increasingly important in the industry as technology, infrastructure, machine learning, and data streams continue to grow in importance, as organizations increase their reliance on data and insights for growth and cost reductions. It will show how TML leads to frictionless machine learning by reducing human touchpoints in the machine learning process and elastic by quickly scaling up or down TML solutions. It will also discuss how the TML solution approach can handle and analyze, by machine learning algorithms, any amount of data volume. It also shows how algorithms can be managed to reduce unnecessary cloud compute and storage costs. The role of governance and security is also discussed to support a nearly closed-loop approach to the machine learning process. It ends with concluding remarks.

Chapter 3 will present some of the industry challenges with data streams and machine learning – specifically, why it's still a challenge to not only use and apply conventional machine learning to data, but more so with streaming data. As industries continue to leverage data and find ways to reduce costs while trying to find ways to engage more with their customers, there exist both human and technological challenges that prevent many organizations from fully embracing machine learning to help guide every area of their business. These challenges have led to a growth of third-party technology and consulting vendors that specialize in data streams and data science to help organizations to digitally transform their businesses and embrace data and machine learning to stay ahead of their competition. This chapter will also discuss the approaches to addressing these challenges and presents a path forward for companies that struggle with fully embracing data streams and machine learning. It ends with concluding remarks.

Chapter 4 will focus on the business value of TML. Rarely do technology books focus on how the technology will provide value to businesses. In this chapter, we will focus on several areas of value: people, processes, technology, and culture. Each of these value levers will be discussed within the context of TML. We will also discuss the risks of TML, discussing how it is not a silver bullet, rather part of the overall technology toolset that can be applied to many industry use cases. It will discuss how TML changes the conventional machine learning processes by reducing some of the friction points. These friction points will be identified and compared with TML and without TML. The chapter concludes with closing remarks.

Chapter 5 discusses the technical components of a TML solution with data streams and AutoML. This chapter will present the individual technologies and how these technologies integrate with each other. It will also discuss the computer code that binds these technologies together. The intention is to show you how a TML solution comes to life as a functioning solution. We will also discuss how the TML components advance the area of data streams and AutoML that will allow users to apply TML to real-world use cases. The chapter will also discuss the pros and cons of the technical components and architecture for organizations. This chapter will set up the subsequent chapter on the application of TML to a real-world business use case.

Chapter 6 will discuss the business applications of TML. It will develop a TML solution template that you can use for any type of industry problem with data streams. This solution template will provide a step-by-step guide that will make it easy for you to start building TML solutions with real code, using the cloud as the data backbone, and technologies for AutoML and visualization that you can download. We will discuss how TML solutions can be technically architected. The chapter further discusses the business value of TML for businesses. The chapter concludes with closing remarks.

Chapter 7 will discuss visualization. It will discuss and show how to visualize streaming insights from TML models. It will show how TML visualizations can be used for decision-making. It will discuss visualization from the Walmart and fraud examples that will help you build your own visualizations with TML solutions quickly and effectively.

Chapter 8 will discuss the future path for TML. The focus here is to provide a vision on where the TML industry will be in 3–5 years and what will the key areas be to drive growth in the field. The intention is to give a consolidated view of data streams and AutoML that make up TML solutions. It will also focus on frictionless, and elastic, machine learning and the areas in the industry that will benefit and those areas that may not. It will discuss the risks of not using TML when it makes sense to do so and the value of cloud technology and infrastructure-as-a-service business models and how this will fuel the adoption of TML. The chapter will also provide a framework to help organizations decide which use cases are better suited for TML and which ones are not. It will further discuss how businesses will integrate TML into their digital organizations to continuously reduce the friction from machine learning processes. It will also discuss the financial value of TML and how it can help to reduce costs and generate new revenue streams for organizations. The chapter ends with concluding remarks.

Chapter 9 will provide a summary of the book and the key points discussed in the chapters. It will make clear the core connection between data streams and AutoML, which when combined create TML solutions. It will focus on how TML leads to frictionless, and elastic, machine learning solutions and how this can add business value fast. It will also discuss how TML projects are managed and estimated.

This book ends with a list of references used throughout to support our claims and assertions from using research from others. The book also adds information about the authors and finally concludes with a section on acknowledgments.

CHAPTER 2

Transactional Machine Learning

Transactional machine learning (TML) is an exciting area of machine learning that is specifically focused on applying auto machine learning (AutoML) to data streams to create machine learning solutions that are frictionless and elastic. This chapter will define TML and the related technologies that are important to build TML solutions. It also discusses the (big) data platform that will store and process the data streams for analysis and machine learning. Applying machine learning to data streams will be vital for disruptive technologies [Jayanthiladevi et al., 2018]. We define TML as follows.

TML Definition: The ability of a computer to learn from data streams by using automated machine learning applied to the entire, or partial, data stream set that leads to a frictionless, and elastic, machine learning process and solution that is continuous and mostly uninterrupted by humans.

Examining TML

Let's take a closer look at what we mean by TML. Transactions are data produced or created in real time or data that are generated and captured right after some event. For example, you produce transactional data when you engage in a certain activity like buying a product or service: a point of sale (POS) machine captures and records data generated from your actions like how much you paid for the product or service, quantity bought, your name, product or service name, purchase location, payment information, and so on. These transactions represent the latest event of buying a product or service and therefore capture your current behavior or action. As data speeds increase, the opportunities to make faster decisions also increase. TML enables faster decision-making that is critical for time-sensitive business use cases such as fraud

© Sebastian Maurice 2021
S. Maurice, *Transactional Machine Learning with Data Streams and AutoML*,
https://doi.org/10.1007/978-1-4842-7023-3_2

detection, product and service recommendations, IoT device control, and so on. Table 2-1 shows the evolution from data mining and business intelligence to TML. As data speeds increase, we are evolving from relational data warehouses to event streaming platforms, which will require higher automation of machine learning processes that will enable faster decision-making. ML solutions, for select business use cases that require a higher degree of human interactions for ML model building, testing, and deployment and create higher friction, will evolve toward lower friction as more ML processes are automated. Furthermore, lower friction in ML solutions will also lead to higher elasticity of ML solutions that can scale up or down based on the needs of the business by adjusting the number, and components, of ML solutions quickly.

Table 2-1. *Evolution to TML*[1]

	Data Mining and Business Intelligence	**Batch Machine Learning**	**Transactional Machine Learning (TML)**
Data Refresh Rate	Days	Hours, minutes	Seconds, milliseconds
Decision Speed	Reactive – human-in-the-loop decisions	Hybrid human/machine loop	Instant – AutoML loop
Data Environment	• SQL data warehouse • Data mining middle layer • Data visualization front end	• Data lake storage • Cluster computing framework • Machine learning infrastructure	• Data stream infrastructure • Event-driven streaming cloud platforms • AutoML service • Microservices-based data architecture

(*continued*)

---

[1] I would like to thank Tim Raiswell for the key ideas and points in this table.

Table 2-1. (*continued*)

	Data Mining and Business Intelligence	Batch Machine Learning	Transactional Machine Learning (TML)
Use cases	Low-volume decision support: • Business performance diagnosis • Customer profitability analysis	Medium-volume decision-making: • Ad placement • Product and service recommendations • Dynamic pricing	High-volume decision-making: • Fraudulent transaction identification and intervention • Product and service recommendations • IoT device controls
Machine learning solutions	• High-friction ML solutions • Low elastic ML solutions	• Medium-friction ML solutions • Low-medium elastic ML solutions	• Low-friction ML solutions • High elastic ML solutions

Machines can also create transactions of data by sensors that measure the electronic signals of the machine. Internet of Things (IoT) connected devices such as TVs, cars, appliances, industrial machinery, and other similar devices are a good example. In the IoT context, without human interactions, machines are more independent to produce data. This is possible because of advanced sensors or other electronic circuitry that allows a machine to produce data which are measured by the sensor; this measurement gets stored in a database or file systems. For example, some refrigerators can *sense* when it is running out of milk and then alert a human to buy milk. Or, the refrigerator can place an order to buy the milk itself. While autonomous machines have advanced in the past years mainly due to the lowering costs of sensors[2] which have fueled the IoT market, the use of sensor data for machine learning is still in its infancy but growing.

Adding intelligence to machines has several advantages in three areas:

1) Decision optimization

2) Risk reduction

3) Cost savings

---

[2]https://new.engineering.com/story/research-report-how-the-shrinking-cost-of-sensors-is-fueling-the-internet-of-things

First is in the area of decision optimization. You try to make optimal decisions, in most part, using information available to you. Even your decision on what to eat tonight hinges on what you feel like eating, your location or proximity to available restaurants or access to apps, what is in your refrigerators, what others feel like eating, how much money you have, what your likes and dislikes are, and so on. You may also use machine output like the ratings of restaurants on apps in your mobile phone, even what restaurants are recommended by the GPS software in your car.

Machine decision-making, and optimization, is also possible and becoming important in the IoT industry. Machines can use data to perform health checks on themselves. For example, predictive asset maintenance (PAM) use cases can be designed to allow machines to make a service call to a human to come fix it [Faiz and Edirisinghe, 2009]. Decision-making by machines reduces the need for humans to monitor machines for potential failure. By allowing machines to use their own data, it reduces the friction caused by humans and can dramatically reduce any adverse impacts that machine failure can have on people and businesses. But the challenge is *how* do you get a machine to think independently to make data-driven decisions using data streams? TML provides a way of how this can be done, and it will be discussed in this section.

Second is risk reduction. If a machine were able to decide by itself when it is about to fail, could this help to preempt machine failure? Certainly. The ability to predict events before the events occur allows you to plan for potential machine failure. We say potential, or likelihood, because this is not yet certain. For the machine to compute the probability or likelihood of failure, it will require the machine to learn the patterns in the data to produce a probability measure. But learning the patterns in the data requires the machine to use algorithms to understand the patterns. Recall the weather temperature example where we constructed a simple model of using cloud cover (CC) and time of year (ToY) to help predict what the weather temperature (WT) will be tomorrow. We showed, very simply, how the machine learns the patterns and correlations between all the variables to produce the estimated coefficients: a, b, c values. Note that this was a multidimensional problem with three dimensions: WT, CC, and ToY. This is where the value is realized by machine learning: the ability to construct a multidimensional relationship between the dependent and independent variables that data mining or stream mining cannot fully capture.

Building complex relationships, through machine learning models, unearths more information on how the independent variables affect the dependent variable. Therefore, for the machine to predict the likelihood of machine failure, it will likely need to construct a multidimensional model that, after its coefficients are estimated, can be used to predict the probability of failure.[3] The family of machine learning models that can predict a probability of an event occurring are classification models such as logistic regression, decision trees, and random forest models [Faiz and Edirisinghe, 2009].

Third is in the area of cost savings. Creating financial value from TML is an important aspect that makes it attractive for businesses. Machine learning labor costs are reduced due to less human touchpoints in building, large, machine learning solutions. Data management costs are reduced since TML does not require data schemas. TML employs schema on read, which makes TML solutions flexible and not dependent on any particular schema and expensive data warehouses that have to be maintained. Using a cloud service–based approach, TML solutions are elastic and can be quickly customized for diverse business needs and users. This agility helps to create an environment that is more likely to provide benefits than incur costs. For organizations, this benefit should result in an increase in profits from their operations by increasing revenues and reducing costs. There are many areas that TML can be used to increase profits across different industries, and this will be the focus of Chapter 4.

Features of TML

Low friction and high elasticity characterize the operational aspects of TML solutions. These operational aspects drive lower costs and higher benefits for organizations. These benefits are a direct result of the features of TML solutions. There are five key features as follows:

1) Data is fluid – Just like streams of water, data flows in from any direction (source) and flows out to any location (sink).

2) Data streams must be joinable – Just like SQL[4] joins, joining data streams is critical to building a training dataset in real time.

---

[3]We ignore for the moment the discrete classification of the dependent variable that identifies when in the past the machine has failed and when it has not.

[4]SQL stands for Structured Query Language. Please see www.iso.org/standard/63555.html for more information.

3) Data stream format is standardized – Data streams are standardized to JSON format. If data streams are standardized in JSON, this makes it easier to build portable solutions between systems and technologies and perform analysis.

4) Data streams are integrated with auto machine learning (AutoML) – AutoML applies machine learning, without human intervention, to training datasets to find an algorithm that best fits the data streams.

5) Low code – TML solutions provide coding languages that make it easy to develop TML solutions without the need to write extensive lines of code.

The preceding features will be important to help you classify use cases that can be addressed by TML. These features can also provide guidance to solution architects in architecting TML solutions that will require specific people skills, streaming technology platforms and machine learning process changes to be built into the architecture. These features should not be seen as a complete list; it may be that over time new features are added as technology evolves and business requirements change. We now discuss the preceding features in detail.

Data Fluidity

Data streams, as already discussed, are characterized by their velocity, variety, veracity, and volume. Data fluidity simply means that the source of these data streams can come from anywhere or anything: this is what we mean by source. The streams of data can also go anywhere: this is what we call sink. The concept of **source** and **sink** will play an important part in developing TML solutions that are able to scale easily to address large business use cases.

Within a TML solution context, we can further translate source and sink into publisher of data and subscriber to the data, respectively; this can be called a **pub-sub** model [Eugster et al., 2003]. Data in a data stream has to be generated by someone or something and consumed by someone or something, that is, humans or machines. For TML solutions, publishing has two components:

1) Publishing raw data to the stream

2) Publishing the insights extracted from the raw data using AutoML into another stream

Subscribers can subscribe to the insights in the stream. This ability to publish or produce data to data streams and then subscribe to consume the insights from another stream will be a core part of a TML solution and fundamental to reducing the friction from machine learning processes. The mechanics of how this is done is discussed in Chapter 5.

Figure 2-1 shows the producer and consumer process of a TML solution. Process step 1 shows the producers of data. These could be any data sources generated by anyone or anything. To process these data, we need a middleware software that ingests the data and may additionally transform the data.

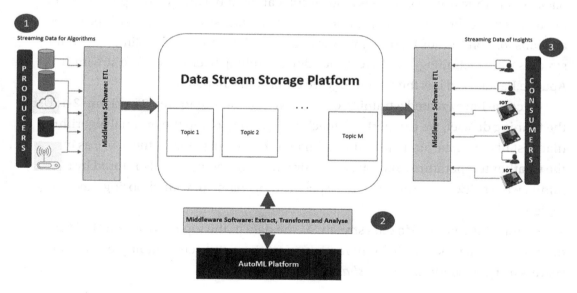

Figure 2-1. *Transactional Machine Learning Process*

This middleware is an important component of a TML solution. Its main function is to ingest and store it in the data stream storage platform (DSSP). The data stream storage platform must be able to handle very large amounts of data because a key characteristic of data streams is *volume*. The volume of data could range to thousands of terabytes or more. The platform also needs to ensure the data stored can be easily accessed. There are technologies available to handle massive amounts of data [Jayanthi, 2016]. In addition to storing large amounts of data, the DSSP needs to perform another important function: it needs to have the ability to **roll back** in the data stream and retrieve historical data. Rolling back the data stream is important to construct training datasets for machine learning. Specifically, we already know that data in streams accumulate very fast; the accumulation of data is simply data being stacked on top of each other or **appended** to the data stream. Each byte of data increases the storage capacity. Imagine 1 byte of data being generated by ten data sources, and stored every millisecond, 24 hours a day. In one day, you will need 864,000,000 bytes of storage or 864,000 kilobytes and that is just for 1 byte of data being generated. Big data storage technologies use specific algorithms to break up the data into smaller sizes and store it in distributed computer systems [Jayanthi, 2016].

If we did not have the ability to roll back the stream, we could not build a training dataset for TML. Having the ability to extract a **sliding window** dataset, by rolling it back, in real time from a stream is a powerful feature. It becomes more powerful when extracting these datasets from a continuous flow of data streams because it provides a means for continuous learning. Very few data platforms allow for rolling back data streams. One technology to pioneer the ability of rolling back data in the stream is Apache Kafka,[5] and it is the technology we employ in Chapter 6.

Once we have created a training dataset, we can then go to step 2 in Figure 2-1. Using the training dataset, we can perform machine learning to find the optimal algorithm that best fits the data, along with the estimated coefficients, that is, the a, b, c as in the weather temperature example. The estimated parameters are also stored by the middleware software in the DSSP in a specific topic name of your choosing, say, insight topic 1.

A topic is the name of a data stream. As the data in the topic grows, the DSSP can distribute the data across different memory partitions and across multiple computer servers using a technique called *sharding*.[6]

---

[5]Apache Kafka, `https://kafka.apache.org/`

[6]Sharding is a technique for breaking up large amounts of data and distributing them across the same machine or different machines that can be later retrieved.

Step 3 in Figure 2-1 shows consumers subscribing to the topic and consuming the insights in the topic and visualizing the streaming results.[7] A subscription to the topic creates a dedicated connection to the topic, or data stream, for consumption. A consumer could be a human or a machine. The consumption of the insights can be used for reports and dashboards or sent to downstream systems. Note that in steps 1–3 we do not interact with any humans, except of course if the consumer is you. In this way, we can design a completely automated machine learning process that is frictionless.

It is beneficial to highlight the core terminology that will be used in later chapters as shown in Table 2-2.

Table 2-2. *TML Process Terminology*

Terminology	Description
Producers	Producers are data generators like humans or machines. Sensors on IoT devices are a good example of producers.
Consumers	Consumers are humans or machines that subscribe to a topic and receive data or insights that can be used for reports and dashboards or sent to other downstream systems.
Topics	Topics are names of data streams. A user creates a topic in the DSSP by giving it a name that uniquely identifies this data stream.
Subscribing	A machine or human can subscribe to a topic if they want to consume data or insights from the topic.
Consuming	To consume from a topic, consumers use the middleware software to first subscribe to the topic, then use the middleware software again to receive data from the topic.
Producing	To produce to the topic, producers use the middleware software to first create a topic in the DSSP, then use the middleware software to send data to the topic in the DSSP.

(*continued*)

---

[7]We will discuss MAADS-VIPERviz, a visualization technology specifically built for visualizing, securely, streaming results from conventional web browsers over dedicated WebSockets using HTTP or HTTPS protocols.

Table 2-2. (*continued*)

Terminology	Description
DSSP	A DSSP is a big data stream storage platform that facilitates a publisher-subscriber model. The DSSP has two key functions for TML: 1) Partitions 2) Offsets Partition is a number that allows the DSSP to manage big data storage efficiently in a distributed network by breaking up the data using sharding and storing it in different partitions across different machines in the network of computers. A partition starts at zero. Users can specify the number of partitions in the DSSP computer server they want. A large number of partitions can improve the accessibility performance for consumers. For example, if 1000 consumers want to consume from the same topic, a partition setting of 1000 allows each of the 1000 consumers to access the topic in parallel without impacting other consumers. Offset is a zero-based number that keeps track of each datum in the data stream. An offset starts at zero to signify the first data point in the data stream and grows sequentially as data in the stream grows. An offset is determined and maintained by the DSSP. The offset will be important because it allows us to "roll back" the data in the stream to refer to past data. By rolling back data in the stream, we can construct a training dataset in real time. A data value in the data stream is uniquely identified in the topic by a partition, offset, and key.
Middleware software (MWS)	A software that allows producers and consumers to send data to the DSSP and receive data from the DSSP. The MWS can be called a connector for the DSSP that performs a microservice. It is also integrated with AutoML software. The MWS as shown in Figure 2-1 connects the **Producer-AutoML-Consumer** together, allowing developers to create TML solutions. MWS will be described in detail later.
Key	A key is a unique value that identifies each datum value in the topic. The purpose of a key helps the DSSP to efficiently manage data across partitions in a distributed network while also allowing consumers to retrieve the information quickly.

(*continued*)

Table 2-2. (*continued*)

Terminology	Description
Training dataset	A training dataset is created by joining data streams, from a particular offset value, that has one dependent variable topic and one or more independent variable topics. The consolidated dataset is called the training dataset and will be used for TML.

Next, we discuss the process of joining data streams.

Joining Data Streams

The importance of joining data streams is fundamental for creating training datasets for TML. As discussed in Table 2-2, each data stream is named as a topic; you can use these topic names to identify data streams you want to join by using the MWS. Each stream is standardized by having the same data format: we use JSON[8] format for each stream. JSON is a popular and widely used format using a key-value notation. For example, this is a simple JSON: {"Name": "Sebastian Maurice", "Phone Number": 555-555-5555}, where "Name" and "Phone Number" are keys, and "Sebastian Maurice" and 555-555-5555 are values, respectively.

To join streams with MWS, you follow these steps:

1) Identify the streams to join by using the topic names.

2) Use the MWS to join the streams.

3) Create another topic to store the joined streams or training dataset.

4) Produce data to the training data stream.

There are some challenges when joining streams. First, individual streams may be generated at different times. This means that joining streams on the creation date and time of the data could be challenging. But this challenge is reduced (if not eliminated) because creating sliding training datasets by joining multiple streams grabs data from each stream at virtually the same time. This assumes, however, that each stream is producing data and not idle. For example, using offset=-1 tells MWS to get data from

---

[8]www.json.org/json-en.html

the end of all streams that can then be rolled back by a user-specified offset to create a training dataset – this will become clearer in Chapter 6. By using the MWS, you can specify any rollback offset you wish to control the size of training datasets.

Second, the length of the streams is likely to differ. In order to perform machine learning on the consolidated data streams, the lengths of the streams must be equal. To create equal lengths, TML employs a pruning algorithm as follows:

1) Grab data from each stream from all partitions using the rollback offset.

2) Due to multiple partitions in each stream, the size of the data will vary. For example, if one stream has 1500 data points across all partitions and another stream has 500 data points across all partitions, the lengths are obviously different. To align the lengths, we must prune one of the streams so that the lengths are the same across the streams; in this case, we prune the longer stream with 1500 data points to 500.

3) Once all the streams are equal in length, join them and create a training dataset.

The preceding process happens in near real time and is a critical step before performing TML.

Third, to consolidate the streams to create a training dataset, we must ensure one stream is the dependent variable and other streams are independent variables; this will be our model for TML. Next, we discuss how standardization of data can simplify joining.

Data Stream Standardization

We use the JSON format for each stream data. The formatting to JSON is automatically handled by the MWS when it produces data to the stream topic in the DSSP. The JSON format follows a generally accepted convention that allows other applications and systems to read the JSON data in the same way, making it easier to analyze the data even if it originated from different sources. In addition to ease of access, it also helps to improve the performance of reading data. The JSON keys act as indexes that can be used to quickly access data which makes referencing each component of the JSON object easier for reporting, analysis, and creating the training dataset. This leads us to the next section of how this integrates with AutoML.

Data Stream Integration with AutoML

A key function of TML is the use of the training dataset, from data streams, for machine learning. The creation of the training dataset therefore represents an important requirement for TML. We use the MWS to connect to the AutoML technology which connects to the DSSP, allowing the AutoML technology to consume the training dataset and apply machine learning to find the optimal algorithm: there is no training data movement between the MWS and AutoML. The AutoML technology then produces this optimal algorithm to a topic in the DSSP, as shown in Figure 2-1. So, the AutoML technology is both a consumer of the training dataset from the DSSP and a producer of the optimal algorithm to the DSSP. The complete AutoML process is shown in Figure 2-2. The figure shows the general process. Let's assume we want to detect credit card fraud with this process and that Topic 1 contains the training dataset of credit card transactions after MWS has joined the relevant data streams to create a training dataset then:

1) MWS consumes from Topic 1 that contains the credit card training dataset.

2) MWS connects to AutoML technology that performs machine learning to find the optimal machine learning algorithm that best fits the credit card transaction data for the best trained model.

3) AutoML produces the optimal trained model to another Topic 2: Credit Card Fraud Trained Model.

4) MWS consumes the optimal trained model from Topic 2.

5) MWS reads new credit card transaction data from Topic 3 and generates insights of potential fraud using the new transaction data. The new data are the independent variable values used to predict the dependent variable (Fraud). For example, in the weather temperature example, Topic 3 would contain cloud cover and time of year values. The prediction is then stored back into another topic, Topic 4: Fraud Predictions.

6) Consumers who want the insights would subscribe to Topic 4 and consume the insights. Specifically, customers can be automatically alerted via email or text message of potential fraudulent transactions.

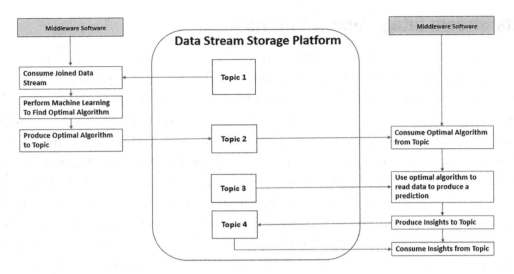

Figure 2-2. *AutoML Process*

The preceding steps are repeated for any number of data streams and can easily scale to any number of algorithms and the insights consumed by any number of consumers. How this is architected will be explained in Chapter 5 on architecture. Specifically, the MWS and AutoML technologies can be used as **microservices** that can be instantiated to any number, easily deployed, and used to scale TML solutions to almost any number of data streams, algorithms, consumers, and producers. Easily integrating AutoML into this scalable architecture is one of the unique aspects of a TML solution proposed in this book. This takes us to the last feature of low code, which discusses how you can connect to the MWS using the MAADS Python library or using REST API, making the MWS language agnostic.

Low Code

TML solutions that are integrated with the DSSP, MWS, and AutoML require programming code for greater control for execution and debugging purposes. However, the amount of code needed should not be excessive and should be as small as possible.[9] Compared to no-code, low code trades off visibility and customization for agility, scale, and speed to insights. While low code is possible using graphical user interfaces, this is

---

[9]wwwcdn.spanishpoint.ie/wp-content/uploads/2019/04/The-Forrester-Wave%E2%84%A2_-
Low-Code-Development-Platforms-For-ADD-Professionals-Q1-2019.pdf

38

not required as prebuilt functions can provide the same amount of value. For example, libraries that provide an API (application programming interface) can alleviate writing code from scratch, thereby reducing the effort it takes to develop TML solutions. In this book, we use a library called MAADSTML Python library[10] to develop TML solutions as shown in Figure 2-3. This library is developed in the Python programming language[11] and can be easily installed in any environment by invoking "pip install maadstml".

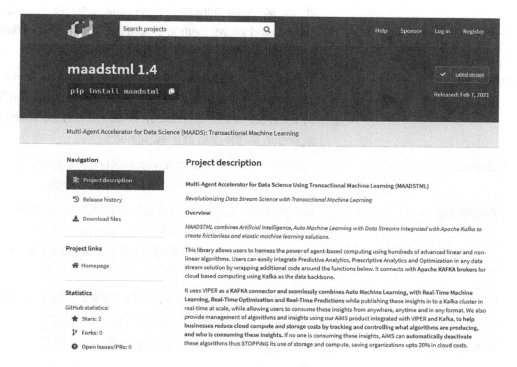

Figure 2-3. *MAADSTML Python Library*

This library contains functions that provide an API to help produce data, process the data using AutoML, and consume data. The core functions in this library are shown in Table 2-3.

---

[10]https://pypi.org/project/maadstml/
[11]www.python.org/

Table 2-3. *MAADSTML Python Library Functions*

MAADSTML Python Functions	Description
vipercreatetopic	Create topics in Kafka brokers.
vipersubscribeconsumer	Subscribe consumers to topics. Consumers will immediately receive insights from topics. This also gives administrators of TML solutions more control over who is consuming the insights and allows them to ensure any issues are resolved quickly in case something happens to the algorithms.
viperhpdetraining	Users can perform TML on the data in Kafka topics. This is very powerful and useful for "transactional learnings" on the fly using our HPDE[12] technology. HPDE will find the optimal algorithm for the data in a few minutes.
viperhpdepredict	Using the optimal algorithm, users can do real-time predictions from streaming data into Kafka topics.
viperhpdeoptimize	Users can even do optimization to MINIMIZE or MAXIMIZE the optimal algorithm to find the BEST values for the independent variables that will minimize or maximize the objective function.
Viperproducetotopic	Users can produce to any topics by ingesting from any data sources.
viperconsumefromtopic	Users can consume from any topic and use it for visualization.
viperconsumefromstreamtopic	Users can consume from a multiple stream of topics at once.
vipercreateconsumergroup	TML administrators can create a consumer group made up of any number of consumers. You can add as many partitions for the group in the Kafka broker as well as specify the replication factor to ensure high availability and no disruption to users who consume insights from the topics.

(continued)

---

[12]We discuss HPDE in Chapter 5.

Table 2-3. (*continued*)

MAADSTML Python Functions	Description
viperconsumergroupconsumefromtopic	Users who are part of the consumer group can consume from multiple partitions of a topic concurrently.
viperproducetotopicstream	Consolidate the data from the streams and produce to the topic.
Vipercreatejointopicstreams	Users can join multiple topic streams and produce the combined results to another topic.
vipercreatetrainingdata	Users can create a training dataset from the topic streams for TML.
viperanomalytrain	Performs unsupervised learning on data streams for anomaly detection.
viperanomalypredict	Predicts likely anomalies from data streams using peer groups found using the viperanomalytrain function.

By using the preceding functions, you can dramatically reduce the amount of code needed to develop TML solutions, as we will show in Chapter 6. Specifically, the MAADSTML functions needed to develop the functionality in Figure 2-2 would be

1) **vipercreatetopic** – This would create Topics 1–4.

2) **vipercreatejointopicstreams** – This would create the metadata structure for the training dataset and starts the training dataset construction process.

3) **viperproducetotopicstream** – This physically grabs the data from the streams using the metadata structure.

4) **vipercreatetrainingdata** – This creates the training dataset using the data from the previous function and produces it to Topic 1: Credit Card Training Dataset.

5) **viperconsumefromtopic** – This consumes the training dataset from Topic 1.

6) **viperanomalytrain** – Using the training data, this performs AutoML and produces the optimal trained model (peer groups in this case) to Topic 2: Credit Card Fraud Trained Model.

7) **viperanomalypredict** – This consumes the optimal trained model (peer groups) from Topic 2 and uses data from Topic 3: New Transaction Data to make a fraud prediction.

8) **viperproducetotopic** – This produces the fraud predictions to Topic 4.

9) **vipersubscribeconsumer** – Consumers subscribe to Topic 4.

10) **viperconsumefromtopic** – This consumes the insights from Topic 4.

An application of TML to a business solution template will be the focus of Chapter 6. Up to this point, we have discussed the process of integrating data streams with AutoML to create TML solutions. We now discuss the DSSP, AutoML, and MWS technologies in more detail.

Data Stream Storage Platform (DSSP)

We have discussed how the DSSP, an event streaming platform, manages the streams of data that we use for TML. For this book, we will use Apache Kafka.[13] Apache Kafka (or just Kafka) is a distributed open source technology used by 80% of the Fortune 100 companies.[14] It was developed by a group of developers at LinkedIn[15] who donated it to The Apache Software Foundation.[16] The developers of Kafka then started a company called Confluent[17] to provide Kafka managed services; as of 2019, the company is worth $2.5 billion.[18] The main reason for Kafka's popularity, aside from being open source, is that Kafka is the only publisher-subscriber event streaming platform [Jayanthi, 2016], which makes it ideal for TML in the following ways:

1) Kafka is able to handle massive amounts of data and manage it effectively across multiple partitions in topics or data streams.

---

[13]https://kafka.apache.org/

[14]https://kafka.apache.org/

[15]www.Linkedin.com

[16]www.apache.org/

[17]www.confluent.io/

[18]www.forbes.com/sites/alexkonrad/2019/01/23/open-source-unicorn-confluent-reaches-25-billion-valuation-three-years-after-hiring-its-first-sales-rep/?sh=70f4ee8715e2

2) Kafka is an event streaming platform optimized for scale, which is different from other big data platforms like Hadoop which are optimized for large static datasets amenable for large batch computing.

3) Kafka uses offsets when storing data in the streams, which allows us to "roll back" data in data streams to create training datasets. As explained in the Kafka documentation,[19] "offset acts as a unique identifier of a record within that partition, and also denotes the position of the consumer in the partition."

4) Kafka allows us to name data streams as topics.

5) Kafka allows us to join data streams.

6) Kafka can run on any cloud platform.

7) Kafka allows groups to consume from one topic, thus allowing for parallel delivery of information to each consumer.

Kafka was originally developed for large messaging systems that benefit from a publisher-subscriber type of framework. However, given Kafka's use of **offsets**, it allows us to extend the functionality of Kafka to machine learning using data streams to create TML solutions. This extension of Kafka to transactional machine learning is the first of its kind currently. We will use MAADS-VIPER as our MWS connector to Kafka: MAADS-VIPER is currently the only Kafka connector for TML that has been verified and tested by Confluent.[20] MAADS-VIPER is discussed next.

MAADS-VIPER

An official Kafka connector, shown in Figure 2-4, has been verified and tested by Confluent. MAADS-VIPER (or just VIPER) is our MWS that facilitates communication between consumer and producers, AutoML, and DSSP. VIPER is core to TML. There are two ways to communicate with VIPER to build TML solutions:

1) MAADSTML Python library using the functions described in Table 2-3.

---

[19]https://kafka.apache.org/22/javadoc/org/apache/kafka/clients/consumer/KafkaConsumer.html

[20]www.confluent.io/hub/oticsinc/maads-viper

2) Representational State Transfer (REST) API is a type of software architecture with defined constraints that allows for the creation of web services. For example, browsing to a website uses the hypertext transfer protocol (HTTP); with this same HTTP protocol, users can connect to VIPER. This ensures that developers, no matter what programming language they are using, can access VIPER.

VIPER fits into a microservices architecture because it

- Is easily deployed to any operating system, that is, Windows, MacOS, Linux

- Communicates over HTTP and programming language agnostic

- Uses SSL/TLS encryption of data to and from the DSSP

- Performs tasks that are focused around specific capabilities of TML solutions

- Is decoupled from DSSP and AutoML

- Can be downloaded from a central source

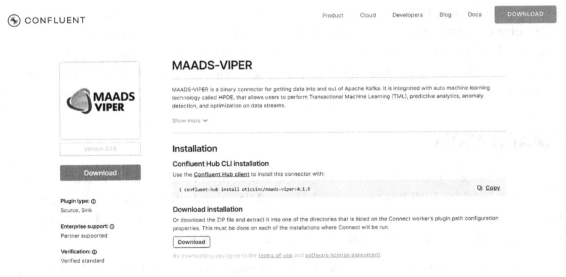

Figure 2-4. *MAADS-VIPER Kafka Connector*

The special functionality, along with easy access to functions, makes VIPER an effective Kafka connector that has been extensively tested and verified by Apache Kafka creators at Confluent.

There are several aspects of VIPER that make it an effective TML enabling technology:

1) VIPER is cross-platform; it can run in Windows, MacOS, and Linux.

2) VIPER is cloud agnostic. For big data solutions, cloud platforms are important. VIPER can connect to any major cloud platform that is running Apache Kafka such as Microsoft Azure, Amazon AWS, Google Cloud Platform, or Confluent Cloud.

3) VIPER is SSL/TLS enabled. Security is critical these days, and organizations want to ensure all data is encrypted. Kafka can be set up for SSL/TLS encryption; this means VIPER needs to handle SSL/TLS traffic to communicate with Kafka.

4) VIPER can read from multiple partitions in a topic concurrently using consumer groups. As data in streams increases, and the number of consumers grow, Kafka will distribute the data in a topic across multiple partitions. This will mean that consuming data from a topic across multiple partitions becomes more challenging from an application perspective. VIPER has a built-in process to consume from multiple partitions so data are always consumed properly and concurrently.

5) VIPER can read from any offset. As mentioned, specifying offsets is fundamental in creating training datasets. However, given the nature of data streams, offsets can play an important role in controlling the volume of data in the streams to construct training datasets that are not too large. If training datasets are too large, this can drastically increase the amount of time it takes for machine learning and will impact performance of the TML solution.

6) VIPER can be instantiated multiple times. For very large TML solutions with thousands or millions of data streams, and thousands of algorithms, it may be required to have many instances of VIPER for load shedding that fits into a microservices architecture. As will be discussed in Chapter 5, using VIPER as microservices is an important architectural consideration for large TML solutions.

The next section discusses how algorithms are managed using the Algorithm and Insights Management System (AiMS) Dashboard.

Algorithm and Insights Management System (AiMS) Dashboard

VIPER can manage a large amount of algorithms via the Algorithm and Insights Management System (AiMS) Dashboard shown in Figure 2-5. A consequence of TML is the relative ease in creating algorithms with AutoML. This will lead to many algorithms that make up a TML solution in circulation. These solutions, therefore, must be monitored and tracked for consumption because they will have cost implications for storage, compute, and network access. AiMS offers a very convenient way to track, monitor, and control the execution of TML solutions.

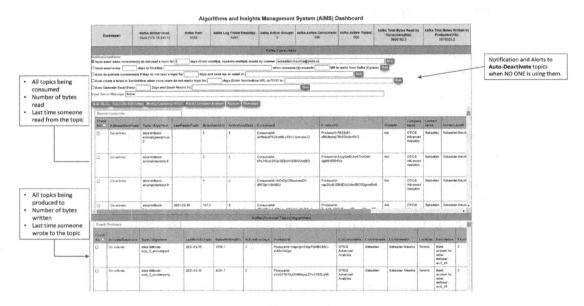

Figure 2-5. *AiMS Dashboard*

The information in each instance of VIPER is saved in an embedded database using JSON format; AiMS uses this information to populate its dashboard. This means that VIPER does not require an external database storage. The advantage of an embedded database is that it makes VIPER portable and self-contained. This ensures that regardless of which system VIPER is installed in, the information captured can be used to manage

TML solutions. Table 2-4 shows the information captured by VIPER for every consumer, producer, and group of consumers. Note that this is not the DSSP; rather this is metadata that VIPER stores in its local embedded database instance for AiMS.

Table 2-4. *AiMS Dashboard Field Names*

Field Names	Description
Activate/Deactivate	Administrators of VIPER can manually activate or deactivate a topic, or they can tell VIPER to automatically deactivate algorithms using the alerts and notifications.
Active Read Days	This is the number of days the topic has been read.
Active Write Days	The number of days that the topic was written to.
Bytes Read (Kb)	The number of bytes, in kilobytes, read by the consumer is recorded. This is the Egress amount. Egress is an important metric for pricing cloud consumption.
Bytes Written (Kb)	The number of bytes, in kilobytes, written by the producer. This is the Ingress amount. Ingress is an important metric for pricing cloud consumption.
Company Name	This is the name of the company the consumer belongs to.
Consumerid	Consumer ID uniquely identifies the consumer that is consuming from this topic.
Contact Email	This is the email of the consumer.
Contact Name	This is the name of the consumer.
CreatedOn	This is the date and time when the topic was created.
Dependent Variable	If this topic is a joined topic that stores a training dataset, then this is the name of the dependent variable in the training dataset.
Description	This is a description of the topic.
Groupid	Group ID is the ID of the group this consumer may belong to. Administrators can create groups and assign consumers to this group. This could be useful for parallel delivery of information from a single topic to multiple consumers.
Group Name	This is the name of the group.

(continued)

Table 2-4. (*continued*)

Field Names	Description
Independent Variables	If this topic is a joined topic that stores a training dataset, then this is the name of the independent variables in the training dataset.
isActive	This indicates if a topic is active (1) or not active (0).
Joined Topics	This will list the names of the topics that are joined together.
Last Offset	When a consumer reads from a topic, this field records the last offset of the consumer in the topic.
Last Read of Topic	This is the last read date and time for the topic by a consumer.
Last Write to Topic	This is the last date and time the topic was written to by the producer.
Location	This is the location where the consumer resides in.
MAADS algorithm key	This is the key name that uniquely identifies the optimal algorithm.
MAADS-HPDE Algo Server	This is the physical IP address or network name of the AutoML server.
MAADS-HPDE Microservice	If using a reverse proxy, this is the network name for the proxy. Using microservices can be very beneficial for large-scale TML solution deployments.
MAADS-HPDE Port	This is a network port that the MAADS server is listening on.
MAADSTML Token	This is a secure token needed to access MAADS-HPDE.
Number of Consumers	The number of consumers in the group.
Partitions	This is the number of partitions in the topic.
Producerid	Producer ID is the ID of the producer that is writing to this topic.
Replication Factor	The replication factor is a feature of Apache Kafka and signifies the number of physical computer servers to use for redundancy.
Topic/Algorithm	The Topic name in the DSSP. This can also be the name of the algorithm and is chosen by the user.
Updated	This is the last date and time the topic was written to by the producer.

Additional functionality for administrators are

- Subscribe consumers to the topic
- Save alerts
- Modify consumer details
- Batch consumer activate/deactivate
- Activate a group of topics
- Deactivate a group of topics
- Create Topics
- Batch topics activate
- Create groups
- Batch activate/deactivate groups

Managing algorithms is important for organizations for three reasons:

a. Cost – Algorithms will consume network computing and storage resources, which will cost money, especially when using a cloud platform. By keeping track of the producers and consumers of algorithms, VIPER makes it easy to determine who is consuming and who is not consuming insights. If no one is consuming insights, then the algorithm is not serving any value but still consuming compute and storage that cost money; in these cases of no consumers, VIPER can notify the administrator of the TML solution and automatically deactivate algorithms that are not being used, potentially saving the organization money.

b. Security – By tracking algorithms, organizations know exactly what the algorithm is doing, who is consuming it, and how many bytes are being read and written. This allows them to control access to the information and only give users access to the information that they need.

c. Service management – Maintaining a good level of service in the event a consumer stops receiving the insights due to a problem in the solution is another important aspect of VIPER. If a consumer complains that the algorithm is no longer producing the insights, the administrator of the solution can immediately look up the algorithm topic in the AiMS Dashboard and determine who developed it and contact that person to fix the issue or create a service management incident.

AiMS offers organizations a bird's-eye view of all the algorithms running in their environment. They have access to create alerts and notifications, shown in Figure 2-6, to control access and allow them to make the best use of resources to reduce unnecessary costs.

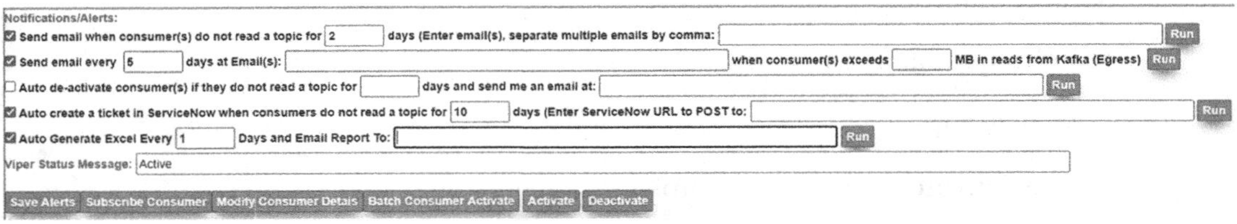

Figure 2-6. *AiMS Notifications*

There are five types of alerts and notifications in AiMS:

1) Send email when a consumer or producer does not read or write to a topic for X number of days.

2) Egress – Send an email when a consumer reads X number of megabytes from a topic.

3) Ingress – Send an email when a producer writes X number of megabytes to a topic.

4) Auto-deactivate a topic when a consumer or producer does not read or write to a topic for X number of days.

5) Auto-create a ticket in a service management software when a consumer or producer does not read or write, respectively, to a topic.

6) Generate, and email, an Excel report listing all of the consumers, producers, and groups.

Next, we discuss the AutoML technology used by the TML solution to perform real-time machine learning.

AutoML Technology

The AutoML technology used by the TML solution is called MAADS-HPDE.[21] It is a cross-platform technology that can run in Windows or Linux. It is integrated with MAADS-VIPER, AiMS, and Apache Kafka and specifically designed for fast machine learning. There are several key aspects to any AutoML technology such as [He et al., 2020]

1) Data preparation

2) Model generation

3) Hyperparameter optimization

4) Model estimation

5) Model reporting

6) Model deployment

MAADS-HPDE performs the preceding six functions. Specifically

- Data preparation – HPDE can read from a comma-separated values (CSV) file, or it can directly connect to an event stream platform like Apache Kafka and directly consume a training dataset from a topic.

- Model generation – MAADS-HPDE applies several algorithms to the training dataset. These include

  - Neural networks

  - SVM/random forest

  - Classification model where dependent variable is discrete

  - Gradient descent algorithms

  - Linear and nonlinear regressions

  - Unsupervised learning for anomaly detection in data streams

---

[21]This technology is developed by OTICS Advanced Analytics Inc. (www.otics.ca) and can be downloaded from https://github.com/smaurice101/MAADS-HPDE. HPDE stands for Hyper-Predictions for Edge DEvices.

51

- Hyperparameter optimization – HPDE employs a randomized grid search process to find the best parameters for algorithms using MAPE (mean absolute percentage error) for continuous dependent variable and ROC/AUC metric for discrete dependent variable. Specifically, HPDE applies several algorithms and computes the MAPE or ROC/AUC for each and then chooses the algorithm with the lowest MAPE or highest ROC/AUC as the **optimal algorithm**.

- Model estimation – It performs model estimation to find the estimated coefficients that best fit the data. This is possible using the **viperhpdetraining** function discussed in Table 2-3.

- Model deployment – It can deploy the optimal algorithm to any server or write the results to a topic.

There are several key areas that make HPDE unique among other AutoML technologies:

- HPDE is integrated with an event streaming platform like Apache Kafka, AiMS, and MAADS-VIPER for real-time modeling. HPDE can deploy the optimal trained models (algorithms) to a topic in Apache Kafka. This makes it easy for other applications that have access to the topic to retrieve the optimal trained model and make predictions.

- Continuous learning – An important aspect of AutoML technologies should be their ability to perform continuous learning by remembering old data and learning from new data [He et al., 2020]. Using the sliding window techniques, TML solutions have zero model obsolescence by learning new patterns in fast flowing data on a continuous basis without human intervention.

- HPDE also does mathematical optimization. Recall that optimization is the process of finding the best values of the independent variables that minimize or maximize some objective function such as maximizing total realized profit by selecting the optimal prices and product mixes on a retail website. This is possible using the **viperhpdeoptimize** function discussed in Table 2-3.

- HPDE performs real-time unsupervised learning (discussed later) on data streams and **does not** require a classification of the dependent variable beforehand – as required by conventional machine learning processes to predict fraud or anomalies. VIPER uses **viperanomalytrain** and **viperanomalypredict** for anomaly detection.

- HPDE is SSL/TLS enabled and encrypts all data written to, and read from, Apache Kafka.

- HPDE can be instantiated to any number of instances and makes it effective for large-scale TML solutions that are distributed across networks.

The AutoML process is shown in Figure 2-7. The process shows the integration between the MAADS-VIPER and MAADS-HPDE. Specifically

- MAADS-VIPER creates the training dataset topic in Apache Kafka.

- MAADS-HPDE reads the training dataset topic.

- MAADS-HPDE performs any data prepping and transformation.

- MAADS-HPDE applies machine learning algorithms to the training dataset.

- MAADS-HPDE chooses the algorithm that best fits the data: we call this the optimal trained model.

- MAADS-HPDE then deploys this algorithm by writing the algorithm location to another topic in Kafka.

This process can be repeated an unlimited number of times, with unlimited number of data streams and training datasets. The only limitation is on the hardware side. Theoretically, with an unlimited amount of storage, the AutoML process can be run with an unlimited amount of data.

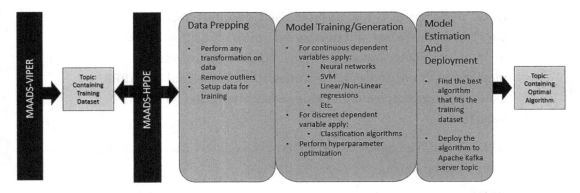

Figure 2-7. *AutoML Process*

Many of the models built using TML will use supervised learning; however, in many other cases, unsupervised learning will be important for anomaly detection in data streams. The next section will discuss how TML can handle fraud or anomaly detection use cases.

Unsupervised Learning: Detecting Anomalies

Unsupervised learning is useful in many cases when there is a need to detect anomalies in the data. For example, use cases for fraud detection normally require the detection of unusual behavior. To determine "unusual" behavior, we must compare it to "usual" or "normal" behavior. This comparison, when using **supervised learning**, is straightforward: it requires past knowledge of "unusual" and "usual" behaviors that are physically classified by a human; then a machine learning model is formulated with these classifications, and a machine learns the patterns of unusual and usual behaviors. However, with data streams, it is almost never possible to know beforehand what is "unusual" and "usual" behavior. Fast flowing data makes it difficult to formulate a supervised machine learning model to allow the machine to learn the differences in behaviors. Lack of past knowledge of anomalies in real time, and speed of data flow, makes the application of supervised learning difficult if not impossible; therefore, we must use **unsupervised** learning with data streams.

HPDE performs unsupervised learning using a variant of techniques developed by [Bolton & Hand, 1999] using peer group analysis (PGA) and break-point analysis (BPA). The technique is reasonably simple as follows:

1) From the data, find a peer group that is similar to the types of transactions to evaluate for fraud or anomalies.

2) Create a summary statistic of this peer group that can be used to measure the differences in "usual" and "unusual" behaviors.

3) Compare the new transactions to this peer group and determine how "unusual" the transactions are to the peer group summary statistic. Flag those transactions that vary greatly from the summary statistic as suspicious.

The preceding steps can get complicated quickly when data are streaming. But, this complication is greatly simplified by VIPER and HPDE with Kafka. Specifically, as shown in Figure 2-8, there are two elements to the unsupervised learning process:

1) Selection of peer groups

2) Anomaly detection to flag behavior that differs from the peer group

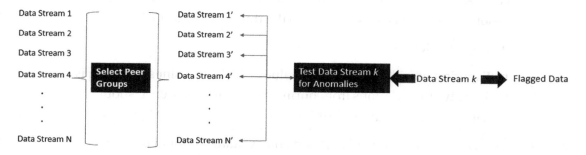

Figure 2-8. *Method 1: Unsupervised Learning Flow*

The selection of the peer groups from data streams involves the following steps:

1) For each data stream, select the data window from the data by simply using the offset parameter to roll back the data.

2) For each data stream window, compute the centroid of the group using a K-means clustering algorithm.

3) For each data in the window, compare it to the centroid of the group to determine the **distance** between each point and the centroid.

4) Store the distances in an array.

5) Sort the array from highest to lowest distance values.

6) Remove the first and last members in the distance array – we remove the first and last members to standardize on values to remove the influence of extreme behaviors and construct the peer group data streams: Data Stream 1', Data Stream 2', Data Stream 3', Data Stream 4'... Data Stream N'.

7) Now compare new data to the peer group, and for each new data, compute the standardized variance metric called T for each new point. We can interpret T as a **score** – the higher the score, the more suspicious the data. A threshold can be set that compares T to the threshold; any T values above the threshold value are flagged as suspicious. A question one could ask is how do you choose a good threshold? The T score for the new data (T^d) is compared against the T score of the centroid value (T^c); the absolute percentage is compared to the threshold value (T^t):

   a. Data suspicious if $(|\ T^d - T^c|/\ T^d) \geq T^t$

   b. Experiment with the threshold value (T^t). If the business use case is sensitive to deviations from normal behavior (T^c), then T^t should be a low percentage around 30%; if not sensitive, then set it higher.

8) Construct the Flagged Data stream and evaluate for fraud or anomalies.

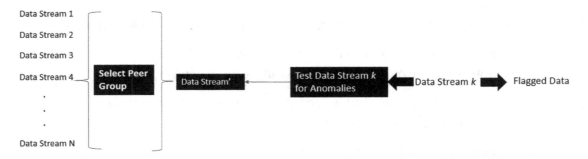

Figure 2-9. *Method 2: Unsupervised Learning*

The break-point analysis (BPA) method is similar to the peer group analysis (PGA) method, the difference being BPA creates **one group** composed of the centroids from all the other groups. This method picks out the usual behaviors from all the peer groups and creates a group with just these behaviors. Another difference between PGA and BPA is the former compares the new behavior across the peers of varying behaviors because normal behavior in one peer group may not be normal in another group: it accounts for the variability of normalness. Using both PGA and BPA allows for a more robust detection of unusual behaviors or behavior that is not unusual. Next, we show how TML leads to a frictionless machine learning process.

Frictionless Machine Learning

Conventional machine learning processes differ from TML processes by minimizing the human touchpoints. Too many human touchpoints lead to higher levels of friction for select use cases. This is evident in conventional machine learning processes where human and machine interactions are frequent in data preparation, model generation, and model deployment processes. This is not to say that human and computer interaction is bad; in fact, for some cases that are highly complex, this can be quite effective in determining the optimal trained model (algorithm) [Amershi et al., 2014]. However, in dealing with data streams, conventional methods are not ideal for reasons discussed earlier. Specifically, data streams are not static disk resident data, rather data that is continuously flowing. This makes conventional machine learning processes difficult to apply by humans.

We can classify the causes of friction in machine learning processes in five areas:

1) Data preparation – Reducing human touchpoints in the data preparation can dramatically reduce friction.

2) Model generation – Automating the selection of algorithms applied to the data can reduce the time humans need to decide which model to choose. As well, setting up the model can be time-consuming.

3) Hyperparameter optimization – Choosing hyperparameters is probably the most time-consuming process. By automating this process, we can reduce friction.

4) Model reporting – Documenting the details of how the model was chosen can also be very time-consuming. Automating model reporting of algorithm results not only simplifies this process but also improves model transparency.

5) Model deployment – Deploying models or algorithms require certain processes be followed, but these processes can be automated. For example, if the requirement is to have the model deployed to a test environment first and then, after a few days of testing, deployed to production environment, this can be automated.

It is important to note that **not** all business use cases fit within a TML process. For example, for highly complex and low data volume use cases, conventional machine learning methods may require human experts to formulate the machine learning model over several iterations and testing scenarios before deploying it for production. The choice of method will depend on many factors. However, in cases when data streams exist and it is possible to join streams of data to create training datasets, then the TML process should be considered.

Furthermore, the TML process still requires that humans choose the right data streams to join, which will be driven by the business use case, and test the results or outcomes from the models before using it in a real-world setting. While we realize that TML cannot completely eliminate the friction that exists in conventional machine learning processes, it can, if applied to the right use cases, considerably reduce the friction by reducing the human touchpoints. In addition, TML is more effective for large, distributed, use cases that have many consumers. TML for small use cases may be overkill; this will be discussed in Chapter 4 on the business value of TML. We now provide concluding remarks.

Concluding Remarks

This chapter has provided an in-depth explanation of TML. It has been formally defined. The most important aspect of TML is its integration of data streams and AutoML technology: DSSP and MAADS-HPDE. It uses MAADS-VIPER as middleware software (MWS) to manage the entire process of (training) dataset creation, producing to topics, consuming from topics, and controlling the entire algorithm development, creation, and deployment process while maintaining access controls, with AiMS. TML technologies broadly cover the end-to-end TML process to reduce friction in the machine learning process.

The ability to create TML solutions with low-code technology like the MAADSTML Python library, in some sense, democratizes the TML process and allows users with minimal development experience to build powerful solutions. Using the MAADSTML functions, shown in Table 2-3, allows organizations to integrate the output of TML solutions to downstream systems. For example, the functions *viperhpdepredict*, *viperanomalypredict*, and *viperhpdeoptimize* can output predictions and optimal results to other topics that can be read by other applications. These applications can use these results in dashboards or reports for human, or machine, consumption.

One of the issues with conventional machine learning processes has been the level of friction that normally exists in data preparation, model selection, hyperparameter optimization, model reporting, and model deployment. These are areas that can add weeks or months to a machine learning solution. The conventional process is also not sustainable for data streams that are continuously flowing. Conventional machine learning solutions are also not elastic solutions that can be scaled up or down or turned off quickly when no longer needed. In these cases, TML may be a better option for data that are constantly changing. Integrating data streams with AutoML allows us to reduce the friction by automating many of the conventional machine learning processes, resulting in dramatic reductions in the number of days it takes to create and deploy machine learning solutions. While not every business use case will fit within a TML process, those use cases that have data streams which can be joined to create a training dataset, and are large scale, may make more sense for TML. The next chapter will discuss the industry challenges with data streams and machine learning.

CHAPTER 3

Overcoming Challenges to ML Adoption

Organizations have many data and ML challenges. Overcoming the data challenge with a solid data strategy, and creating a data culture, is an important step toward applying and adopting ML in your organization. ML adoption will take hold in your organization if it can solve problems that can help your organization grow. While changing an organization's culture is never easy or quick, showing tangible ML value that can be related to cost decreases or revenue increases will drive more ML adoption.

As data evolves, so will ML. Data has already evolved from small to big data, from structured to unstructured, from static to continuous flowing in data streams. While there are varying differences in the degree of these evolutions, they are happening and will continue to accelerate. As organizations face cost and market pressures, and more sophisticated customers' needs, finding ways to automate processes and provide faster and more higher quality information will be important. In this chapter, we will look at how you can build a stronger, more data-driven, organization that embraces ML for deeper insights that is focused on solving key business problems. By using data and machine learning to make data-driven business decisions, organizations can be more proactive in dealing with potential business challenges such as targeting products and services to different customer segments, reducing fraud, pricing products and services more competitively, reducing inefficiencies in the supply chain, and so on. Being proactive in the decisions you make by using insights from data through machine learning will lead to better management of your business by allowing you to adapt quickly to changes in your customers' needs and preferences. The next section provides an overview of the challenges organizations face with data and machine learning today and will face tomorrow.

© Sebastian Maurice 2021
S. Maurice, *Transactional Machine Learning with Data Streams and AutoML*,
https://doi.org/10.1007/978-1-4842-7023-3_3

Overview of Challenges

In a recent article,[1] organizations are still early in their adoption of advanced technologies like AI and machine learning. US Census data from a 2018 survey of 583,000 companies in the United States, and compiled by the National Bureau of Economic Research [Zolas et al., 2020], showed that only 2.8% of companies are using machine learning or 16,324 companies.

Much of the use of advanced technologies is skewed toward large companies with 250 employees or more, which is more than three times the rate for smaller companies with 10 employees or less. Since many small businesses account for the highest numbers of total employees, if small businesses close shop, this could have a large ripple effect on employment and development of new ideas.

Applications of TML for small businesses will largely be in financial services, that is, fraud detection in FinTechs, digital or social media areas where analyzing social media data is important for sentiment analysis or product recommendations. There are applications in manufacturing for predictive asset maintenance using sensor data. Retail areas where social media data is used to determine product or service prices for different segments of the customer base, that is, male vs. female and location differences. Large businesses have similar use cases but at a much higher scale. TML will scale to more data streams, more algorithms, and faster information delivery to different consumers. Data centralization will be important to maintain, but this is also another challenge for organizations and discussed later. Next, we look at the root causes of adopting advanced technologies.

Understanding the Root Causes of Challenges in Adopting Advanced Technologies

We discussed several key challenges that may lead to a slow adoption of advanced technologies. This section attempts to get to the root causes of these challenges so they can be addressed effectively. There are five areas that can hinder the adoption of advanced technologies, such as data streams and machine learning (along with others):

1. Data decentralization

2. Lack of corporate strategy promoting advanced technologies

---

[1] www.wired.com/story/ai-why-not-more-businesses-use/

3. Costs for

    a. Advanced people skills

    b. Advanced technologies

4. Lack of compelling use cases for machine learning

5. The human and corporate will to change to a digital world

These areas will each be discussed in the following.

Data Decentralization

Many organizations still have silos of information that reside in many databases, flat files, and other sources. Decentralized information creates several challenges:

1. Lack of interoperability between different sources of information and applications, which creates overlaps in information creation and use. Many of us have experienced receiving the same information from a company multiple times – this is bad data management. A Gartner report[2] found that organizations that avoid master data management (MDM) risk a 25% reduction in potential revenue gains.

2. Reduced speed to insights from the data, due to lack of interoperability, slows down the decision-making process.

3. Reduced number of insights that can be generated from all data reduce the diversity of decisions that can be made.

4. Duplication in data, resulting from the lack of centralized systems of record, can result in wasted time and effort to curate and maintain data.

5. Reduced quality of decisions, from not using the right data for the decisions, can impede business growth.

---

[2]www.gartner.com/en/newsroom/press-releases/2014-02-20-gartner-says-master-data-management-is-critical-to-crm-optimization

Data silos are bad for business.[3] Lack of centralization, strict access levels, and poor data quality can result in slower innovation causing companies to lose their competitive advantage. Companies that lack a good data strategy that aligns with their corporate goals and objectives risk losing business opportunities to rivals. To understand why data silos exist in the first place, you need to ask these questions:

1) Why are we generating data?

2) Who or what is generating these data?

3) Where is the data being used, by whom, and for what purpose?

4) How can we make better use of our data?

5) What are our major data-generating processes (hint: finance and operations are often major data sources)?

6) How do we manage business-level and corporate function technology purchases and ongoing use?

7) Is there a corporate IT standard for purchasing and managing data-generating systems?

8) Do we have any data governance practices or policies in place to set standards for data creation, storage, and access?

9) Whose role is it to oversee the company's data?

10) Are there incentives for people to share their data or analysis?

Answers to the preceding questions will set the baseline for why you have data in your business in the first place. To understand why there are silos, one only needs to ask:

- Is there value in integrating different data across my business?

If the answer is no, then you are not looking hard enough at the areas of potential value created by combining different data, or maybe there is just no value – but the likelihood of this is very low.[4] For example, integrating data for new hires could help HR from filling out unnecessary onboarding forms. Streamlined payroll processes

---

[3] www.forbes.com/sites/forbestechcouncil/2018/11/19/why-data-silos-are-bad-for-business/?sh=7f8956955faf

[4] www.forbes.com/sites/forbestechcouncil/2018/05/15/the-growing-importance-of-data-integration-between-departments/?sh=639a7a75315c

are another example of integrating HR data with payroll data for both employees and contractors. An integrated data also simplifies access controls and user identity management across business units and jurisdictions. Simply asking the question of how we create business value for areas of the business that are struggling would help generate a conversation about data decentralization and whether it makes sense to combine all the data in a central location without inhibiting innovation [Maurice et al., 2006]. Creating a data culture is important to creating an innovation culture. There are generally accepted steps in creating a data culture,[5] and it starts with your business leaders. However, shifting the mindset of your employees not used to using data for decision-making to then start using it will be a daunting challenge [Waller, 2020]. Creating incentives for employees to use data for decision-making by making it more accessible, removing unnecessary barriers for data access, and getting management to request data-driven reports and analysis will help. The shift to a data culture will not happen overnight, nor can it be mandated; it must evolve together with the business. As data and machine learning show their value through better decision-making outcomes, effective reporting, and improving employee productivity, only then will it become easier to make the shift.

Decentralized data may do harm to not only the bottom line of your business but also negatively impact the credibility and reputation of your business. This damage may not manifest itself immediately, but over time the organization could see a reduction in attracting top talent, increased turnover, reduced sales, increase in negative social media posts, and so on. For some companies, these *signs* will be detected and new defensive strategies created to stem the bleeding. However, many other companies will realize the damage much too late. The companies that detect the negative signs early can take proactive steps to address the deficiencies. Companies that see the early signs are likely using data and insights to inform their decisions to quickly pivot and get ahead of the likely devastating consequences to their business. This leads us to the lack of corporate strategy, discussed next.

Lack of Corporate Strategy

Having a corporate strategy is critical to meet the corporate objectives and goals. TML should be part of this strategy for those organizations where real-time decision-making is important. TML will likely be a part of an innovation strategy, data strategy, skills

---

[5]https://hbr.org/2020/02/10-steps-to-creating-a-data-driven-culture

strategy, and a technology and infrastructure strategy. Many organizations have begun their journeys to digitize their business. For several organizations, *digital* is in their DNA: Facebook, Amazon, Google, IBM, Microsoft, and Apple are the notable ones. These businesses have grown considerably, not organically but by acquisitions. As shown in Table 3-1, in the past 10 years (2010–2019), Google made 37 acquisitions, Apple made 25 acquisitions, IBM made 21 acquisitions, Amazon made 19 acquisitions, and Microsoft made 16 acquisitions [Soni et al., 2020].

Table 3-1. *Acquisitions by Company*

Company	Acquisition
Google	37
Apple	25
IBM	21
Amazon	19
Microsoft	16

However, simply buying AI and machine learning (ML) companies to skip ahead, with AI and ML market innovations, will not work. This is because regardless of how many companies you buy, if you do not have a strategy of how digital technologies are going to improve the core aspects of your business, you will be wasting lots of time and money and will soon realize that you are not any further ahead. There are several high-profile examples of bad acquisitions:[6]

1) Google's acquisition of Nest was worth US$3.2 billion. In an effort to enter the smart home market, Google bought Nest, and the two founders of the company also came over to Google. But, after the acquisition, product innovation slowed, and due to increased political fighting, the two founders of Nest resigned.

2) Microsoft's acquisition of Nokia was worth US$7.9 billion. In an effort to speed up their entry into the mobile phone market, Microsoft purchased Nokia. But Microsoft failed to impress phone app developers, and the infrastructure needed to take on Apple

---

[6]www.cbinsights.com/research/merger-acquisition-corporate-fails/

and Android phones. The acquisition failed, and Microsoft was forced to lay off thousands of employees in Nokia and Microsoft and wrote off US$7.6 billion.

3) Time Warner's acquisition of AOL was worth US$164 billion. The intention was to merge old media with new Internet media. But problems such as lack of cultural fit between the two companies, dot-com bubble collapse, and dial-up Internet going away resulted in a $45 billion write-down in 2003 and a $100 billion yearly loss until AOL was spun off into a different entity.

Having a good strategy will allow you to understand the value you expect to achieve from TML for your organization, customers, and other stakeholders. It will help you to meet timelines for project delivery and allow you to plan changes in the organization that align to ML solution use. This leads us to the costs for advanced technologies as another challenge.

Advanced Technology Costs

Advanced data and machine learning technologies and skills are not cheap. These skills are in demand, and there is no end in sight when this will abate. This creates challenges for organizations when competing for top talent. Combining high labor costs with the costs for advanced technologies can make it difficult for organizations to justify the cost for innovative solutions using ML (or AI[7]), especially if the return on investment (ROI) is not immediately clear. This is one of the main challenges with ML projects: it may take several months, or years, for the returns from ML projects to be realized. This adds a level of risk to the investment that many organizations need to make in ML projects, and in tough times, many are not willing to take on this risk. The level of risk an organization is willing to take should, therefore, be made in conjunction with the right type of information about the ML use case and the problem it is solving. Choosing the right ML use case, with a clearly defined problem statement and value proposition, will be instrumental in balancing risk with potential value. This leads to the next challenge.

---

[7]We will use AI interchangeably with ML while noting that AI is a very broad term that may not necessarily be just ML.

Choosing ML Use Cases

Choosing the right AI use cases is critical for value realization. Nearly 50% of ML (AI) projects fail [McCormick, 2020]. This failure is due to, in most part, choosing the wrong use case without a clear definition of how the solution will add value to the organization over time. Having a framework to choose ML use cases will help to determine which use cases will provide the most value to the organization over time. In addition, not having the right executive sponsorship and misdirected or unrealistic business or technical expectations can contribute to failure. Oversimplifying the complexity of integrating AI solutions into corporate systems, both new and legacy, can impede user adoption and erode trust in the solution. Therefore, organizations should invest the upfront effort to carefully define the goals and problem statements that the AI solution is meant to address or solve.

ML Change Acceptance

Letting go of human decision-making processes and impulses and trusting machines with some of the decision-making tasks is not easy. Nothing is more important than the human will to change. ML is not a normal thing for many people and businesses to use or accept. Distrust of machines, data, an unwillingness to relinquish control to a machine, or all of the above will erode the confidence humans have in these technologies and their abilities to help. For many businesses, it boils down to priorities. We have witnessed unprecedented disruption in major markets with AI-based technologies in almost every industry. The core aspect of these disruptive technologies is not the data or AI; they exploit the human will to accept change when that change benefits them in some way which makes the adoption of AI somewhat easier. If you look at Airbnb, any individual from any walk of life can sell their room or apartment and generate revenue. With Uber, any person with a vehicle and license can offer their vehicle and driving service and generate revenue. With DoorDash, any one with a bicycle or vehicle can offer their own delivery services and generate revenue. All of these companies have great technologies – but if they did not offer something that benefits people and exploits their will to change for something better, they likely would not be worth billions of dollars today. This is similar to organizations: if organizations do not offer their employees or customers something that can benefit them from the change, the willingness to accept change will be close to zero no matter how hard you force the change on them. There are many examples of the internal business value of AI using cognitive engagement with customers, such as chatbots that offer 24/7 customer support for a broad range of queries, internal sites for answering employee questions, product and service recommendations for retailers, health treatment recommendation engines, and so on.[8]

The next section briefly discusses technological barriers – specifically, how technology can be a barrier to change.

Technological Barriers
Skill Gap to Adopting ML

The hypercompetitive space of AI and ML is fueling innovations that are creating new products and services and being embedded into conventional products and solutions such as cars, appliances, TVs, and other daily use equipment and software products. Innovation is an important factor in the growth of economies leading to better products and services, lowering costs for consumers, and creating jobs. But while innovation fuels growth, it needs humans with the proper skills to sustain its growth. Using technology without the proper skills, strategy and processes will not help to sustain continued innovations in products and services. In fact, these barriers may lead to a resentment toward the same technology that was meant to help. This is evident in AI technologies that are meant to make our lives more productive, yet fears of job losses and other uncertainties from AI-based technologies lead to a growing resentment toward these technologies.[9]

Strategy Gap in Adopting ML

The speed of technological innovation is giving rise to speed of data creation. The faster speed of data creation opens up opportunities for faster machine learning. Transactional machine learning (TML) is one of the ways you can leverage this fast data while providing fast machine learning for fast decision-making. But, without a coherent innovation strategy that incorporates data streams, with a recognition for

---

[8]https://hbr.org/2018/01/artificial-intelligence-for-the-real-world
[9]www.forbes.com/sites/cognitiveworld/2018/08/07/job-loss-from-ai-theres-more-to-fear/?sh=6fc8f18523eb

machine learning, an organization will find it difficult to maximize the value from TML solutions. Without an innovation strategy, different parts of the organization can wind up pursuing conflicting priorities [Pisano, 2015]. These conflicting priorities will undermine the overall purpose of technology and its intended uses. While diverse perspectives are critical for growth, without a strategy to integrate these perspectives into an overall, widely accepted, strategy will lead to blunted results that are self-defeating [Pisano, 2015]. TML solutions are simply part of the overall innovation strategy. The components for a successful TML solution will be determined by the types of problems it is solving and how the solution contributes to overall business value, with as little friction as possible.

Communication Gap in Adopting ML

The purpose and goals of technologies must be clearly defined and communicated so it can be clearly understood by the people it will impact. This is no different than a doctor telling their patient that the needle they are about to get will pinch a little. This allows the patient to get physically and emotionally prepared for the needle so there are no surprises. Organizations, in many cases, fail to communicate effectively with employees about the benefits and impacts of technologies they will use in their organization. A good innovation strategy that sets out the purpose, goals, and priorities that are integrated into an overall strategy that is well understood and accepted is a critical step toward achieving adoption of the technology across the organization.

Approaches to Addressing the Challenges

TML solutions require awareness and a desire for speed in decision-making. Speed is central to TML and the reason for auto machine learning. Specifically, fast data creation will lead to fast changes in the patterns and trends in the data. These changes in the underlying patterns and trends will impact the learnings. As patterns and trends change and evolve, so will the learnings that capture these changing and evolving trends and patterns. This speed makes it difficult for humans to build and estimate machine learning models quickly, that is, in seconds. The desire to capture, quickly, the changing trends and patterns requires auto machine learning. This will also require real-time visualization of the insights so that humans or machines can make faster decisions.

Identifying areas in the organization that can benefit from faster decision-making will help to build an innovation strategy that includes TML. Figure 3-1 shows the components needed to help incorporate TML into an overall innovation strategy.

Figure 3-1. *Incorporating TML in Innovation Strategy*

There are five components to consider in Figure 3-1:

1) Central to including TML into an innovation strategy is the awareness and acceptance that speed in data analysis and decision-making is a requirement for creating business value. Analyzing real-time bank or credit card transactions for fraud requires speed in analysis and decision-making. IoT devices require autonomous decision capabilities for self-diagnosis of part failures or health checks. Social media requires fast decision-making for targeted marketing. If there is no requirement for faster data analysis and decision-making, then there is little need for TML.

2) Skills needed for data streams are different than the skills needed for static data. This is due to the following reasons:

    a. Data engineering – Data streams are not the same as columnar dataframes; engineering them requires an understanding of data and events in time.

    b. Feature engineering – Fast changes to the underlying data structure of data streams make conventional feature engineering methods like recursive feature engineering (RFE) difficult to apply. Methods that incorporate the time component to isolate features will be important. For example, the Z-score feature can be applied in real time to detect fraudulent transactions.

71

   c. Data (stream) science – ML on streaming data poses unique
      challenges in sampling, time series modeling, and live model
      performance management that do not exist in batch machine
      learning. Event streaming platforms like Amazon Kinesis, Apache
      Kafka, Apache Flink, Apache Spark, Snowflake, MongoDB, and
      Databricks all require knowledge of event streaming languages
      and processes. Data scientists are more focused on static data,
      whereas data stream scientists are more focused on data streams.
      Specifically, in Kafka, the nomenclature of topic, partitions,
      and offsets are specific to data streams and will be important in
      building TML solutions. Additional areas of knowledge are

      Stream joining required to build training datasets on the fly for
      machine learning.

      Knowledge of machine learning to build training datasets
      composed of a dependent variable stream and independent
      variable streams is critical for supervised learning. Knowledge of
      unsupervised learning is useful for anomaly detection.

      Knowledge of Python programming or a programming
      language is needed to write low-code TML solutions.

      Visualization skills are needed to build big data streaming
      dashboards that display streaming results from the TML models.

   d. MLOps – Conventional operationalization of ML models require several
      human touchpoints in change management, deployment scripts, data
      changes, and so on. These types of processes for data streams are not
      ideal. For TML solutions, models are built and deployed on the fly. They
      can be deployed to a test environment for further vetting and then quickly
      deployed to production.

   e. Model management and governance – Conventional ML model management
      is fragmented and lacks proper governance. It is not uncommon to have
      isolated ML solutions running in an organization servicing a unique purpose.
      With data streams, TML solutions are centrally managed. They are also elastic
      solutions that can be scaled up or down by controlling the number of data
      streams, algorithms, and insights that are consumed by individuals. TML
      solutions can also be activated or deactivated instantly.

3) Business areas that need to make fast decisions need to be identified. Fast decisions could be made by humans or machines. Choose areas that are transaction focused. For example, if you are managing online transactions, then fraud detection could be very useful. If you are trying to reduce cyber-security threats, then analyzing the transaction logs of users to predict likelihood of a threat will be important. If you are trying to improve product quality, then reducing defects in each product by predicting the likelihood of defects will be important for quality controls.

4) Technology choice is an important factor for successful TML solutions. While there are several event streaming platforms, the unique characteristics of Apache Kafka have several advantages:[10]

   a. Offsets – Kafka's offset parameter is critical to roll back streaming datasets to build training datasets with historical data in the streams.

   b. Topic – Kafka names streams as topics. TML uses these topics to join streams to build machine learning models in real time and then performs machine learning on these models.

   c. Partitions – Kafka manages massive amounts of data efficiently by storing data in partitions (within topics). This improves performance, for parallel processing, while building scale in TML solutions.

   d. Replication factor – Kafka replicates data across different Kafka brokers to maintain redundancy in the event of failure in one of the brokers.

   e. Pub/sub – Kafka employs a publisher and subscriber model. This makes it very effective in managing TML solutions and algorithms.

---

[10]A good comparison with another streaming platform, Apache Flink, can be found here: www. confluent.io/blog/apache-flink-apache-kafka-streams-comparison-guideline-users/

5) The type of decisions you make is an important determinant of the success of a TML solution. A TML value framework (TVF) will be developed in Chapter 4 to help you choose TML use cases. Choosing TML use cases rely heavily on time and value. The timing of the decisions and the value of the outcomes of those decisions on the business will help to increase adoption of TML solutions. Choosing the right decisions, with the right business outcomes, will help to build trust and credibility in those solutions that will lead to higher adoption rates and more TML solutions.

Building a proper innovation strategy by considering all of the components outlined will help to ensure that TML solutions are fully integrated into the overall business strategy. Given the highly evolving nature of AI and machine learning, an innovation strategy must also evolve and be flexible. While the components outlined will change and evolve, what should be constant is the willingness by technology leaders to revisit the strategy and make changes as the need arises. The next section will summarize the discussions and outline a path forward.

Discussion and Path Forward

The challenge with data streams and auto machine learning is not only about raising the awareness that speed in data analysis and faster decisions are good for business, but also the willingness to change and accept that data and machine learning are a core part of doing business. Machine learning has separate challenges from data streams such as in data and feature engineering, data science, MLOps, and model management and governance. TML is highly dependent on not only data streams but human expertise in understanding problem domains and data stream science skills that need to translate those business problems into a form that a machine will understand. Then, there are machine learning architect skills that need to architect TML solutions for scale, security, and adoption. The challenges are further complicated by the speed of data creation and desire for faster decision-making. Then there are challenges in system interoperability and data type fragmentation. While these challenges may seem insurmountable, they are not impossible to address. An important aspect of successfully addressing these challenges is to develop a good innovation strategy that identifies the need for faster data analysis and decision-making with wide acceptance from different areas of your business. Clear definitions and communication on the value of fast data analysis for fast decision-making with TML are critical to avoid misunderstandings.

All innovative technologies and solutions are born from a need or desire to improve some product or service that benefits someone or something. Disruptive technologies exploit our willingness to accept change, if that change is a net benefit to us. This benefit improves our lives and businesses. Without some benefit, there will be no wide adoption of the innovation. TML solutions provide a way to analyze fast data for fast decision-making. It is an area of machine learning that is solely focused on speed, value, and scale, with the benefit of providing you with faster ways of making decisions with fast data.

TML also helps create solutions that are frictionless and elastic. By reducing human touchpoints in the machine learning process, businesses can reduce costs, but also increase the speed to insights. Reducing friction, however, does not remove the need to carefully plan out TML solutions by identifying the problem you are trying to solve, then identifying the data streams you need to solve those problems. This requires knowledge of the business domain and also requires data science skills. These applied data (stream) science skills, when coupled with business domain skills, will help to create TML solutions that directly solve the problem.

The elastic nature of TML solutions allow developers and administrators to quickly scale up or down TML solutions as the business needs change. Specifically, elasticity of TML solutions is characterized by three components:

- Number of data streams used

- Number of algorithms created

- Controlling TML solutions by activating and deactivating streams

With TML, you can increase or decrease the number of data streams in a TML solution by joining them and controlling their output. Creating a TML model with 5 or 500 data streams requires the same amount of effort, because all you are doing is passing the number of streams to a MAADSTML Python function to join the streams. The joined streams are used for machine learning. How this is done will be explained in Chapter 6. What is important to realize is the amount of effort involved in building small or large TML solutions is the same, but the cloud storage, network throughput, and compute costs will vary. The other aspect of elasticity is the number of algorithms created. Joining streams is what you will do when you want to create a training dataset for machine learning. AutoML (MAADS-HPDE) will consume the training dataset and find an optimal algorithm. You will use this optimal algorithm for predictive and prescriptive analytics. You can create thousands of algorithms in a very short amount of time. But, this can create model management issues that may result in inefficient use of resources and potential waste of cloud compute and storage resources. But, there is a solution to this. If some TML solutions are no longer in use, developers can immediately, and automatically, deactivate them. Or, if more users want to consume insights from the TML solutions, the developer can simply subscribe more consumers to the TML solution, and they will get immediate access to the results from the TML solutions' algorithms. Elasticity is a unique feature of TML solutions that differentiates them from most conventional machine learning solutions. AutoML is designed to be more elastic than traditional solutions, but with TML you can scale up and down data volumes in a simple UI, and the ML component will adapt with very little or zero human intervention.

So, while there are challenges with data streams and machine learning, TML offers a way forward that allows organizations to meet the demands for fast decision-making with fast data. While not every business problem will need TML, organizations can incorporate it in their innovation strategy and use it when the time is right. The increased awareness that the strategy will create for faster decision-making with fast data will establish the foundation for choices that will need to be made with regard to people skills and resources and technology and encourage a mindset that is needed to meet the increasingly digital world head-on.

CHAPTER 4

The Business Value of Transactional Machine Learning

Fast data requires fast machine learning for fast decision-making. This is the core theme that underpins TML, and it is where the opportunities lie. Specifically, fast learning leads to fast decision-making that leads to faster realization of potential benefits from those decisions. For businesses, fast decision-making can be a source of competitive advantage in a commercial market where you are competing with other businesses for market share and customers [Davenport, 2006]. Speed in learning, by combining data streams with AutoML, is a fundamental departure from conventional machine learning (CML). The reduction in the friction due to the integration of AutoML for online learning and predictions offers opportunities for quicker scaling of solutions. It not only creates a frictionless machine learning process but also an elastic machine learning solution that can scale up or down as the (business) needs change.

As businesses face a post-Covid world, cost pressures will drive the need for increased operational efficiencies and broader adoption of AI.[1] For example, Covid has forced organizations to update, or re-create, demand and sales forecasts several times a week. Why? The new data and, therefore, the models were obsolete in hours. Imagine an always-on model, and process, that just needs new data to reconfigure itself and immediately starts providing updated information: this is the business value of TML. Having the capability of seeing the changes in predictions is a major advantage of an always-on model, because you can see how confident you can be in your business decisions.

---

[1] www.forbes.com/sites/kenrickcai/2020/07/10/ai-50-founders-post-coronavirus-predictions/?sh=696f71442ccc

© Sebastian Maurice 2021
S. Maurice, *Transactional Machine Learning with Data Streams and AutoML*,
https://doi.org/10.1007/978-1-4842-7023-3_4

There is a structural shift happening toward more automation and machine intelligent solutions that is driving a broader adoption of digital technologies that are centered around cloud storage and compute. Speed of analytic insights will become a competitive advantage that will drive business outcomes. The growth in data and increased creation speed of data will demand faster ways of learning from data. Having the ability to learn faster will lead to faster decision-making, which will lead to better business outcomes.[2]

Making faster decisions has some key advantages. It allows humans and machines to adjust their processes or behaviors that result in faster actions. The benefits of making well-informed, data-driven, decisions accumulate over time that further add to improvements in processes or behaviors [Davenport, 2006]. Using data to make good decisions is foundational to improving these processes and behaviors [Davenport, 2006].

For business leaders, data-driven decisions are critical for improving business outcomes [Davenport, 2006]. As we have discussed in previous chapters, TML offers opportunities to learn from fast flowing data that conventional machine learning cannot fully offer. Specifically, by reducing the human touchpoints in the machine learning process, which leads to less friction, organizations can take advantage of speed to insights that allows them to integrate these insights in every business process [Mohammadi et al., 2018]. The value of having data-driven insights to inform every business decision creates

1. An organization that is digitally mature, making it more agile in fast-changing environments

2. An organization that is transparent, which creates more trust in customers and stakeholders

3. An organization that is frictionless, requiring fewer IT FTEs for data, server, and network management and fewer FTEs for data scientists

These three areas of maturity, transparency, and friction can unlock value for businesses that can open new revenue streams that will fuel growth. Digital maturity is critical for success.[3] Using technology to reduce costs is important, but companies are

---

[2]https://hbr.org/2020/01/when-data-creates-competitive-advantage
[3]www.forbes.com/sites/bernardmarr/2020/05/27/digital-maturity-is-critical-to-business-success--now-more-so-than-ever--survey-finds/?sh=3a483b0935fc

now looking to technology to increase growth. Furthermore, transparency coupled with less friction in business processes has a dual effect of increasing trust in your employees and customers. One study found that 94% of consumers would be loyal to a transparent brand.[4] The next sections will take a closer look at how this can be achieved with TML. But first, we discuss conventional machine learning processes.

Conventional Machine Learning (CML)

In this section, we look at how conventional machine learning is focused on static, disk-resident data. This will be helpful to contrast it with TML, which is focused on nonstatic data streams. Today, machine learning is a process that involves several human touchpoints in data preparations, feature selection, model formulation, model estimation, hyperparameter fine-tuning, and model deployment [Amershi et al., 2014; Mitchell 1997]. Human involvement in the machine learning process creates friction that reduces the speed in model formulation, model estimation, hyperparameter fine-tuning, and model deployment; by automating these tasks, we can reduce the friction in conventional machine learning processes. Most data scientists spend 80% of their time cleaning and preparing data, and only 20% doing data analysis.[5] With TML, data analysis is automated, resulting in an immediate 20%, or more, savings. Data prepping with TML involves identifying and joining data streams to build ML models on the fly using the MAADSTML Python library or REST API; this will reduce the data prepping time by an additional 30% or more. We will demonstrate these tasks in Chapter 6.

We are not implying that there is no value in CML; there are many uses for CML that operate and learn from static, disk-resident, data. Some of these uses are focused on decisions that do not need to be made in real time. Or, uses that analyze data that do not change frequently over time. Figure 4-1 shows the CML process within the context of real-time data and scale to make predictions and optimization using optimal algorithms. This figure shows how CML processes, in most part, work offline to analyze data to determine an algorithm that best fits the data. The best algorithm is normally found by choosing and applying several model iterations by changing the parameters, or variables, in each algorithm or applying different types of algorithms.

---

[4]www.forbes.com/sites/mikekappel/2019/04/03/transparency-in-business-5-ways-to-build-trust/?sh=4c6d7a5e6149

[5]www.infoworld.com/article/3228245/the-80-20-data-science-dilemma.html

Choosing the best algorithm is both a science and an art. Statistical metrics are used to gauge goodness of a particular algorithm while avoiding overfitting, meaning avoiding training an algorithm too well, such that it performs well on data it is trained on and not so well on data it is *not* trained on. Once an algorithm and parameters are chosen, they are then used for predictive and prescriptive analytics.

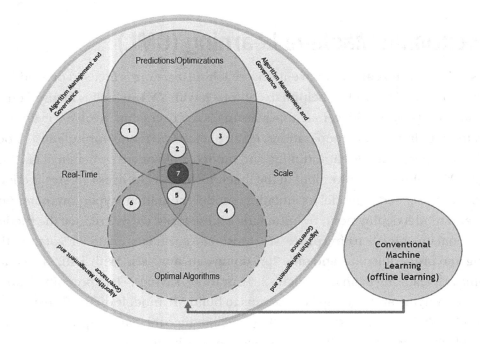

Figure 4-1. *CML Process*

The following explains the six circles in Figure 4-1:

- Real-time – This signifies that data are generated in real time or in batch. These data are not for training the models; rather it is used for scoring or making predictions using the pretrained models.

- Scale – This signifies the enormous amounts of static data and high volume of data across a large distributed, or on-premise, network. Scale in a CML environment is largely dependent on people, process, and technology. Furthermore, scaling a machine learning model involves several components such as server infrastructure, storage, compute, and network throughput. All these will enable thousands or millions of consumers to consume the insights from your machine learning model. Scaling a few machine learning models may not be so difficult, but what if you have hundreds of machine learning models being used by thousands or millions of users across distributed networks? Then, what about model management of these models, security of data and insights, and governance when models need to be changed, retrained, and redeployed? Not easily done with CML. Challenges with model management, governance, and security controls on data and insights make it harder to scale quickly.

- Predictions and optimization – This signifies the use of pretrained machine learning models to make predictions by scoring data and find optimal values of independent variables. This would, for example, be CC and ToY values in our weather temperature example. Specifically, after the model is trained, the estimated parameters (i.e., a, b, and c) capture the learnings from the training dataset. These machine learnings are, in essence, an estimation of the distribution that is generating the data. Recall that the training dataset is a sample of some population data, but we do not have the population data in most cases, so we use the sample data to estimate the population data. It is important to note that these are pretrained models. These pretrained models are developed using static, historical, data that could range over multiple months or years. This is a key departure from TML, where models are trained in real time using data that were generated a few seconds or few minutes ago.

- Optimal algorithm/conventional machine learning – This optimal algorithm is the one that best fits the data, with the best-tuned hyperparameters. It is the one that data scientists, in CML, are looking for and why it could take them days or weeks to find it. This is also a key difference with TML, where optimal algorithms, with tuned hyperparameters, are found in a few seconds or minutes without human intervention. The red arrow signifies the output, from the offline machine learning process, of the optimal algorithm that best fits the training dataset. This optimal algorithm can then be used to make predictions and/or find optimal values for the input variables.

- Algorithm management and governance – Management and governance of trained models is a growing challenge in CML. This is due to three reasons:

  - The growing numbers of trained models being created inside the organization with few standard processes and controls in data access, machine learning platforms, increase in open source technology use, and very little post-model deployment maintenance and support.

  - Organizations and customers are demanding faster insights from the data that improves their day-to-day tasks in some way. This is fueling the need for more machine learning models that is complicating model retraining, versioning, and quality of insights and visualizations.

  - Costs of maintaining and supporting machine learning models will grow as data and machine learning platforms in the cloud, with open source technologies, make it easier to build and deploy models across distributed networks. This will make it harder to manage which models are being used and which ones are not, in an effort to control cloud compute and storage costs. In a recent report by the World Bank for the FinTech industry, many FinTechs reported an increase in expenses for data storage.[6]

TML differs from CML in three areas:

- TML does **online machine learning** on real-time, event-driven, training datasets. Here, online refers to the creation of the training datasets on the fly for machine learning. These training datasets are continuously created as new data comes in.

- TML **uses event streaming data** as both the input data for online machine learning and predictive and prescriptive analytics and optimization. Data streams are populated with events in real time

---

[6]www.worldbank.org/en/news/press-release/2020/12/03/fintech-market-reports-rapid-growth-during-covid-19-pandemic

that require faster response from machine learning solutions to capture the information, process it, and send it onward for fast decision-making by a human or a machine.

- TML solutions are faster to deploy, scale, and retrain: they are **frictionless, and elastic**, which is a key differentiator from CML solutions. CML solutions can also scale, but the cost function is linear. TML's cost function flattens quickly as multiple streams can be added cheaply in the same infrastructure. Elastic TML solutions combine the characteristics and design of TML solutions, with the management and governance of solutions.

The next section looks at the opportunities that TML offers for applying machine learning to real-time, event-driven, data.

The TML Opportunity

The increased growth of data collected by organizations still remains, largely, untapped. In a recent 2019 survey[7] of 64 C-level executives representing very large corporations, such as American Express, Ford Motor, General Electric, General Motors, and Johnson & Johnson, found that

- 72% of the respondents have yet to create a data culture.

- 69% have not created a data-driven organization.

- 53% have not yet treated data as a business asset.

- 52% admit to not competing on data and analytics.

Clearly, the preceding survey shows room for improvement in the use of data and analytics. Of course, on the flip side, 48% of companies are competing on data and analytics. While treating data as an asset is growing, it will require organizational leadership that prioritizes data and insights as a competitive advantage. Moreover, the insights extracted from data will also need to, directly, show how it is contributing to

---

[7]https://hbr.org/2019/02/companies-are-failing-in-their-efforts-to-become-data-driven

increasing revenues and decreasing costs. Without a direct connection to the bottom line of an organization's income statement [Maurice et al., 2006], it will be difficult to gain executive buy-in for using data and machine learning as a competitive weapon. We will show later how organizations can align TML use cases to corporate priorities to ensure the right TML use cases are developed and will directly contribute to these priorities while adding value across the organization.

Figure 4-2 shows a similar diagram as for CML, but here TML is added with real-time data, scale, predictions/optimization which can lead to faster decision-making. We will discuss how organizations can evolve toward TML and add additional value to their organization.

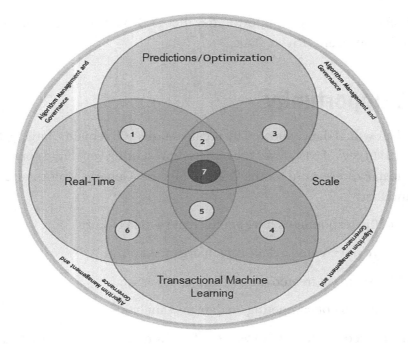

Figure 4-2. *TML Venn Diagram*

The following explains the five circles in Figure 4-2:

- Real-time – This signifies that data are generated in real time and are event-driven; we consider real time as a continuous flow of data such as data streams. These data streams are unbounded, meaning they can technically grow to an unlimited amount. This is a key value area with TML: its capability of performing machine learning on big data quickly.

- Scale – This signifies the enormous amounts of data streams and high volume of data across a large distributed network. With the enormous growth of different types of data from different sources, and organizations' shifting to the cloud, there will be a growing need for cloud-based machine learning solutions that can be built, deployed, managed, and used quickly; managing here means, among other things, having the ability to deactivate solutions automatically if not being used.

- Predictions and optimization – This signifies the use of machine learning algorithms to make predictions and find optimal values of independent variables. TML uses freshly trained machine models to predict or score data. The key difference with CML is that the data used by TML solutions is always the most recent data that can be a few seconds or a few minutes old, not months or years old, as in most CML use cases.

- Transactional machine learning – See the preceding definition.

- Algorithm management and governance – Real-time management of TML solutions to control cloud compute and storage costs by automatically deactivating solutions for lack of use.

Combining the preceding five areas starts to bring to light seven other capabilities of interest:

1) Combining predictions/optimization with real-time data becomes real-time analytics. This is what several companies use today to predict consumer behaviors, recommend products or services, and so on.

   a. Businesses that embed machine intelligence in their products and services offerings have more engaged customers.

   b. Optimization to determine the optimal quantity to sell or buy at a particular price ensures that revenues are maximized and costs are minimized.

2) Real-time analytics at scale allows organizations to predict consumer behaviors and recommend products or services not for hundreds of people, but for millions of people. Consider the billions of people who use Facebook; using real-time analytics at scale allows Facebook to target ads to the individual user. There are a handful of companies that can do this well.

3) Combining scale and predictions/optimization, when not accounting for real time, is the same as applying offline machine learning to static data. Examples here are plenty, such as in health care when predicting the immune response from taking certain drugs. Other examples are predicting product recommendations, retail prices, health and safety incidents, asset failures, and so on. Several companies can do this. Having to update, retrain, and fine-tune the hyperparameters of machine learning models is challenging; TML is a build-it-once approach. It allows organizations to focus less on the inputs that go into TML models and processes and focus more on the outcomes from TML and use it to make better business decisions to drive their business forward quickly. This is important in highly competitive markets.

The areas that are less adopted by organizations (if any) are

4) Combining TML with scale – Very few, if any, companies can do this today.

5) Combining TML with scale and real time, such as performing data mining on streaming data – Very few, if any, companies are doing this today.

6) Combining TML with real time – Very few, if any, companies are doing this today.

This brings us to the core area that remains untapped with TML [Mohammadi et al., 2018]:

7) Combining TML with real time at scale for predictions/ optimization – This is only possible if one combines data streams with AutoML. As we discussed, CML processes are not quick enough to apply machine learning to data streams. Specifically, as data changes, so will the underlying structure of the data, and this will change the learnings from these data. Consider the fast changes in financial data that is driven largely by human behaviors. Using CML to learn from financial data that is months or years old may miss the changes in human behaviors brought on by changing market conditions, environmental changes, pandemics, and the sort. These changes, if missed, will impact the outcomes predicted by machine learning models, which will impact the decisions made by humans. Having the capability to learn faster from the latest data that is easy and seamless has a higher chance of capturing small, or big, changes in the underlying data structure that will naturally lead to better or more accurate model outcomes and decisions.

The outer area of the figure shows the importance of **algorithm governance and management**.[8] Together with value areas 1–6, organizations can evolve toward 7 by focusing on areas that incrementally lead to faster data creation, with faster application of machine learning, for faster optimized and predictive decision-making, with proper governance.

But, what makes area 7 so special? Just because very few companies, if any, are at 7 does not make it desired, or does it? As explained, areas 1–3 are what several organizations are doing today; this already establishes the adoption of real-time predictions and optimization at scale. As more companies begin adopting areas 1–3, this will force leading companies that are already at 1–3 to start to differentiate, innovate, and evolve into other areas to gain market share and remain competitive [Davenport, 2006; Mohammadi et al., 2018].

---

[8]This is the AiMS Dashboard discussed in Chapter 2.

To make fast decisions, fast machine learning (and deep learning) is required with streaming data.[9] But there are several challenges that make it difficult to learn from data streams, namely, reliance on historical data to train an algorithm [Mohammadi et al., 2018]. TML provides a way to extract historical data from data streams to facilitate machine learning quickly. This allows organizations to evolve toward areas 4, 5, and 6 and finally to area 7, which could help them to maintain a competitive advantage.

Evolving toward areas 4, 5, 6, and 7 will require organizations to adopt a different mindset and culture that leverages past innovations in analytics. Davenport (2006) provides several examples of how companies leverage people, processes, technologies, and culture to compete on analytics. Many of these organizations that successfully use analytics build an ecosystem that allows analytics to thrive. For example, Davenport (2006) explains how Walmart insists its suppliers to use their retail link system to monitor product movement by store, plan promotions, and better manage inventories of products using data. Procter & Gamble shares its data and analysis with its suppliers and retail customers to help improve responsiveness and reduce costs [Davenport, 2006]. Building an ecosystem where analytics provides a real purpose and source of value across a distributed network of users with connected tasks to achieve common goals will help to accelerate not only the adoption of data and analytics, but will promote new ideas and solutions that solve new business problems. TML makes data sharing between partners easier by combining or joining your partner's data streams with yours using the cloud infrastructure as the common storage and delivery platform.

A deeper look into how organizations evolve toward analytics shows that it is data, coupled with a business need, that is the driving force [Davenport, 2006]. As more data evolves from low speed to high speed, from low volume to high volume, from low variety to high variety, so will the ways to analyze these data. We only need to look back at the evolution of computers from the Atanasoff-Berry Computer (ABC), Colossus, ENIAC, the Manchester Baby Machine, the EDSAC, IAS, and many more. With the increases in computer memory and storage, powerful (open source) programming languages, and advances in electronic circuitry, we can see how quickly things have changed and

---

[9]Recall that streaming data, or fast data, is constantly changing over time (has temporal locality), so conventional machine learning processes that require humans to operate on mainly static data cannot quickly capture the underlying structure in fast data, whereas AutoML that requires little to no human intervention can be designed to quickly capture the underlying structural changes in fast flowing data.

how we have accelerated toward computing that has opened up many new areas of applications and innovations [Williams, 2000]. Computations and computing would not be possible had it not been for Alan M. Turing and John von Neumann who solved strategic and complex computational problems that drove the need for innovations.

It is not difficult, therefore, to accept that leading organizations will accelerate toward areas 4, 5, 6, and 7 as a natural progression driven mainly by business needs, competitive pressures, technology advancements, and people skills. The next section discusses the core value areas of TML.

Core Areas of Value from TML

Data streams integrated with AutoML that can address large-scale problems, and offer faster insights to millions of people, will continue to gain traction. The previous section has shown how evolution in computing and computations driven by business needs, competitive pressures, technological progress, and people naturally progress to faster, better, and cheaper ways of doing things; this is also true for data and machine learning. The business value from TML, therefore, will be realized through the following areas:

Faster decision-making – By reducing the friction that exists in CML, TML increases speed to insights with little human intervention and more elastic solutions.

Faster scale – Together with Kafka, TML solutions can scale fast using an API and microservices architecture, opening up new opportunities for organizations in almost every industry.

Faster (deep) insights – Joining data streams to create training datasets on the fly, then applying machine learning to find optimal algorithms, with fine-tuned hyperparameters, creates a frictionless machine learning process that can provide deeper insights in a very short period of time, as opposed to days or weeks.

The preceding areas highlight the path to more speed if businesses want to maximize value from fast flowing data. Many researchers [Davenport, 2006; Aggarwal, 2007; Amershi et al., 2014; He et al., 2020; Jayanthi et al., 2016; Read et al., 2020; Yao et al., 2019] have already shown the value of analytics, big data, data streams, and AutoML on decision-making. Recent growth in streaming technologies like Netflix, Apple Music, Disney+, Amazon Prime Video, Facebook streaming, FinTechs, Google, and so on, as well as IoT devices and technologies, shows that consumers' behaviors are moving toward a stream-based culture. These technologies are contributing to an exponential growth in streaming data that will fuel the need for faster ways to analyze these data to make faster decisions. The next section discusses these value areas (levers) in more detail.

TML Value Areas (Levers)

There has been a lack of research in identifying success factors affecting the use of AI on Big Data [Duan et al., 2019]. Even less is the research on the success factors in the use of AI and machine learning on data streams. This section will further discuss the value areas (levers) discussed in the previous section. Specifically, we present here a TML value framework (TVF) to help identify use cases for TML that are aligned to corporate priorities and goals. The main goal of TML is to continuously apply auto machine learning to data streams to extract insights fast: speed to insights with minimal friction is critical for a successful TML solution. Within a business context, there is little to no research to show how AutoML, when applied to data streams, gives rise to business value: increase in revenue and/or decrease in cost. Table 4-1 presents the TVF as a way for organizations to gauge the business value of TML use cases over time.

Table 4-1. TML Value Framework

TML Area	Business Area	Decision Area	Value Generated	Time to Realize Value
Faster decision-making	Identify the areas in your business that require faster decision-making. These could include • Finance • Human resources • Marketing • Information technology • Manufacturing • Risk • Operations	Identify the decision areas. These could be • Increase customer satisfaction • Reduce product failure • Reduce fraud • Increase customer retention • Reduce employee turnover • Improve customer engagement	Identify the expected value[10] you hope to realize for each business area and decision area.	Identify when you want to realize this value. Be specific. For example, *I want to realize an increase in customer retention in 1 week.* Or, I want to realize an increase in customer satisfaction in 1 month.

(continued)

---

[10]"Expected value" here means the value you hope to achieve from likely outcomes. For example, exceeding the average revenue from the past 3 years by 10% could be something to achieve. If decisions are made faster, what might be the average revenue realized or cost avoided? How many such decisions might we make in an hour? A day? A week?

Table 4.1. (*continued*)

TML Area	Business Area	Decision Area	Value Generated	Time to Realize Value
Faster scale	Identify the business areas you want to scale.[11]	What decision area do you want to scale?	What is the value realized from scale? For example, if you want to improve customer engagement for your 10 million customers, then what insights or service can you offer your 10 million customers so they feel more engaged with your product or service?	How quickly do you want to realize this value? If you are providing insights to 10 million customers on product recommendations, are you providing this information on-demand or event-driven, every minute, every hour, etc.?

[11]By scale we mean increasing the number of decisions to be supported by machine learning. For example, you might be a manufacturing company with several hundred machines that throw off data streams. You could make hundreds of local, human optimization decisions about maintenance or output that are locally optimal but collectively suboptimal. But if those machines streamed IoT data to a TML solution, you could develop a single decision framework allowing you to optimize your maintenance and output decisions.

Faster (deep) insights	Identify the business areas that require deeper insights from machine learning: predictions about something or optimal values of something. Do you use models in production currently in a batch or manual update format? Do you find that models lose relevance over time, for example, their performance evaluated by accuracy or error rates, or profitability deteriorates over time? Does your data team spend time updating or refining existing models? If you answered yes to any of these questions, you may have an opportunity to use TML to reduce friction in your data science processes.	Identify what you are predicting or optimizing. For example, I want to predict what my customer satisfaction will be next month. Or, I want to know how many products my customers will want next week. Or I want to detect fraud in every bank transaction.	Identify what the value is from the deeper insights. For example, if you are predicting fraud in 1 million transactions, then indicate what percentage of fraud you want to reduce.	Identify how fast you want to provide these deeper insights. For example, I want to analyze 1 million transactions every second for potential fraud and reduce fraud in 1 month by 10%. CML processes that operate on static data will not uncover potential fraud fast enough and may leave your organization exposed to revenue leaks.

Table 4-1 shows a framework to help organizations choose use cases and business problems that can meet their corporate goals by extracting insights fast. We can show how this framework is applied to a fraud use case for an imaginary company. For this company, we identify the following:

- Business area – Operational risk management in an online retailer selling luxury goods.

- Decision areas – Is a given transaction fraudulent or not, and should the transaction be allowed to continue? At what risk level should a transaction be allowed to continue to checkout?

- Value areas – Your company loses $700,000/week on fraudulent transactions. Reducing this down by 50% prevents $350,000/week in revenue from leaking out of your organization. Now, how would these savings rank against competing use cases?

- Time to realize value – Finance wants to reduce fraud by 50% every week and realize an $8.4 million savings in 6 months.

Being specific, clear, and focused in the objectives will make it easier to design a TML solution that meets your goal(s). The next section takes a closer look at measuring value from TML solutions.

Measuring Value from TML Solutions

The last section discussed the value areas. In this section, we translate and quantify the value generated from TML solutions. Measuring value from TML solutions allows organizations to prioritize and plan TML solutions for development based on highest to lowest value. While a philosophical discussion of value is beyond the scope of this book, a distinction between value and valuation is an important one [Hartman, 1969; Hartman, 1972]. A value of something is largely subjective, dependent on how well it fulfills the concept it was designed to fulfill. Hartman (1972, p. 250) defines value as "A thing has value in the degree that it fulfills the intention of its concept." Valuation is a human activity that measures the properties of the value things – or as Hartman (1969, p. 215] states, "the combinatorial arrangement of the thing's properties." While distinctions between value and valuation may seem subtle and almost trivial, their importance in approaching value calculations is far from trivial, especially if you consider the cost involved in developing large-scale machine learning solutions that could impact millions of people. If the value generated from those solutions is not equal to, or greater than, the cost of the solution, then this is a money-losing solution.

Extending the notions of value and valuation, we can look at value of TML solutions in two ways:

1) Intrinsic value – This is the value the TML solution has in itself; this can relate to the uniqueness of the solution, specifically the uniqueness of the decisions made, the size of the solution, how many data streams are being used, and the number of algorithms generated over time. For example, real-time fraud detection on continuously flowing data transactions that can be done at scale, with lower costs for people and technologies, while reducing the chances for lost revenues would be a value that directly impacts your organization's bottom line.

2) Extrinsic value – This is the value that is not intrinsic. Specifically, what is the external impact of the TML solution to the organization and beyond? How is the TML solution, directly, contributing to increased revenue and/or reduced costs? How long is the payback period with TML solutions?

Understanding both the intrinsic and extrinsic value of TML solutions can help organizations to not only prioritize solutions for planning and development purposes, it also helps to give visibility to solutions across a wider group of technical and business stakeholders; this could accelerate buy-in from stakeholders.

Capturing intrinsic and extrinsic value using widely accepted approaches such as balanced scorecard (BSC) [Kaplan and Norton, 1996, 2001] can help to provide a cross-enterprise view of all TML solutions. And, whether there are opportunities to integrate TML solutions with other non-TML solutions. For example, Table 4-2 shows a BSC [Kaplan and Norton, 1996, 2001; Looy et al., 2016] for TML solutions.

Table 4-2. *TML with Balanced Scorecard*

BSC Area	Observed Perspectives	Performance Indicators	TML Solution Adds To
Financial performance	Financial performance for shareholders and top management	Strategic financial data	• Business area: No • Decision area: No • Value: Yes • Time: No
Customer-related performance	• Customer performance • Supplier performance • Society performance	• Customer survey results • External collaborations and process dependencies • Social media interaction data, i.e., likes/dislikes, negative/positive comments on products and services	• Business area: Yes • Decision area: No • Value: No • Time: No
Internal business process performance	• General process performance • Time-related process performance • Cost-related process performance • Process performance related to internal quality • Flexibility-related process performance	• Descriptive data on process work • Time-related data on process work • Operational financial data • Capability of meeting end user needs • Data on changes or variants in process work	• Business area: Yes • Decision area: No • Value: No • Time: No
Performance related to learning and growth	• Digital innovation performance • Employee performance	• Innovation of processes and projects • Staff contribution to process work and personal development	• Business area: Yes • Decision area: No • Value: No • Time: No

The last column in Table 4-2 called "TML Adds To" should indicate a Yes/No to the TML areas – we filled in the Yes and No as an example. This should be done for every TML solution and will help to tie TML solutions to the overall organization's BSC. This could also be important to promote a non-silo approach across business areas and could make TML solutions more visible to increase buy-in. However, the difficulty arises when some areas of a TML solution add to the BSC area and others that do not. Take the following example from Table 4-2 for financial performance:

- Business area: No

- Decision area: No

- Value: Yes

- Time: No

In this case, Value: Yes means the TML solution adds to financial performance, but other TML areas like decision area and time do not. For instance, you could ask: are some of the decision areas not important for financial performance? Is a No to time implying longer-term performance and not short-term performance? How, when, and by how much will TML reduce fraud? Having the ability to create agile proof of concepts for free[12] can be a quick and risk-free way to evaluate potential TML solutions. We will propose an approach to reconciling this difficulty in the next section. Specifically, the next section will show a valuation approach to quantify the value of TML solutions to help choose TML use cases. The business value provided by TML solutions will be an important aspect of their success.

---

[12]Using the Confluent Cloud platform, users receive $200 free cloud credits that allow them to build TML solutions with enormous amounts of data using TML technologies. This ability to show the value of TML for free in a big data environment is a great way to determine the value of TML solutions quickly and risk-free. If value is validated, the solution can be more formalized as a project.

Choosing the Right TML Use Cases

In some cases, organizations rush into the development of solutions without realizing the broad implications on costs.[13] A careful approach to evaluating the benefits and costs of TML solutions will help to mitigate risks of overspending on solutions that generate very little value for the organization; if a solution costs more than the value it creates, it's definitely not worth developing.

In this section, we develop a TML value index (TVI) that quantifies the intrinsic and extrinsic value of TML solutions. The index components are as follows (within the context of the TML solution):

1) Business area importance

    a. 1 = Very important

    b. 2 = Important

    c. 3 = Slightly important

    d. 4 = Average

    e. 5 = Not important

2) For each decision being made, choose if the decision is

    a. 1 = Very important

    b. 2 = Important

    c. 3 = Slightly important

    d. 4 = Average

    e. 5 = Not important

3) For the value being generated for each decision, choose if the value is

    a. 1 = Very important

    b. 2 = Important

---

[13]www.forbes.com/sites/joemckendrick/2018/02/15/too-much-corporate-money-is-evaporating-into-the-cloud-survey-suggests/?sh=4c72a3731c18

    c.   3 = Slightly important

    d.   4 = Average

    e.   5 = Not important

4) For the time component of the decisions to be answered, indicate whether the time is

    a.   1 = Very important

    b.   2 = Important

    c.   3 = Slightly important

    d.   4 = Average

    e.   5 = Not important

You may ask "Isn't every business area and decision made very important?" Not necessarily. Business priorities change over time. These priorities will give importance to one area over other areas: businesses are not constant. For example, if you are a pharmaceutical company making vaccines and we are in a pandemic, then this company will prioritize their research and development (R&D) more highly than their IT department. So, TML solutions that target the R&D area will, and should, be more important than TML solutions that target IT, within the context of the pandemic. Once the pandemic is over, this business will likely reprioritize their focus areas.

Recall the difficulty of reconciling TML areas within a BSC approach: we can now address this difficulty. Specifically, in cases where TML areas are not uniform in their value add to a BSC area, by using the TVI we can quantify, and rank, the TML use case by their TVI value. Table 4-3 shows how we can calculate the TVI for TML use cases.

Table 4-3. *TVI Calculation*

TML Use Case	Business Area	Decision Area	Value Area	Time	TVI
"Finance area wants to reduce fraud by 10% for 1 million transactions in 1 month"	2	1	1	3	1.75
"Marketing wants to spend 1 million dollars to increase sales of 1000 products by 10% and increase revenue by 4 million dollars in 3 months"	3	2	1	2	2

What is the TVI for this use case? By using the "Very important" to "Not important" scale for each area, organizations could answer this in a group, by taking individuals from each business area as respondents.[14] The TVI will range from 1 to 4: it is a mean of the values of each of the TML areas. Three important points to note:

1) The selection of the importance must be done in the context of the business priority at the time.

2) The selection must not be done in a silo, rather involve as many diverse business stakeholders as possible.

3) To choose among TML use cases, order the TVI column from lowest to highest; the TVI with the lowest number is the most important use case. In Table 4-3, last column, the TVI value of 1.75 is more preferred to TVI value of 2.

While the preceding method works well when looking at individual use cases, it does not work well when looking at the relative importance between use cases.[15] This is because we are converting ordinal ranking to cardinal ranking by assigning a numerical value to importance, which raises concerns on the level of objectivity in the selection of the values for importance. This is why performing this exercise, of choosing values for the TML areas, is important to do in a group between relative use cases. The level of selection bias should be reduced when several diverse individuals make selections [Kugler et al., 2012]. The overall group results should then be used for the TVI calculations. The next section will look at the benefits and costs of TML solutions and how this can be calculated.

---

[14]It is beyond the scope of this book to delve into group decision-making. There exists considerable research in the area [Shaw 1932; Kerr et al., 2004; Kugler et al., 2012; Saaty et al., 2008].

[15]There are methods like Analytical Hierarchy Process (AHP) available to gauge the relative importance of choices in an analytical way [Saaty 1982].

Benefits and Costs

Up to this point, we have focused on areas of value that can be realized with TML solutions, but not much on comparisons of the benefits and costs. Organizations should pay close attention to the benefits generated by TML solutions and the costs incurred to build and support them to avoid wasting financial resources, which is becoming more common.[16] Specifically, the net benefits (i.e., benefits minus costs) should be positive to justify TML solutions. But these net benefits should not be a one-time calculation; rather, net benefits should be an ongoing calculation over the life of the TML solution.

There are several important cost aspects of TML solutions, and some may not be immediately apparent. To see this, we present the following example. We focus on costs for a large TML solution using cloud technologies like disk storage, memory, compute power, and throughput. Each of these components is infrastructure specific to support the operations of TML solutions.[17] Other cost aspects for TML solutions are Kafka specific, as shown in Table 4-4.

Table 4-4. *TML Cost Components[18]*

Cost Type	Amount (USD)	Maximum Limits
Reads	$0.11/GB	100 MB per second
Writes	$0.11/GB	100 MB per second
Storage	$0.00013889/GB-hour	5000 GB
Partitions[19]	$0.004/partition-hour	2048

---

[16]www.forbes.com/sites/joemckendrick/2018/02/15/too-much-corporate-money-is-evaporating-into-the-cloud-survey-suggests/?sh=4c72a3731c18

[17]This will become clearer in Chapter 5 when we discuss the technical components of a TML solution.

[18]These amounts are currently charged by Confluent Cloud.

[19]Partitions have further limits: Ingress up to 5 MB per second, Egress up to 15 MB per second, and storage up to 250 GB.

Now, for a large TML solution, assume

1) Network reads of 1000 GB per hour.

2) Network writes of 1000 GB per hour.

3) Uses 10,000 GB of storage per hour.

4) Creates 10,000 topics/algorithms, each with 10 partitions using 100 GB of storage.

5) TML solution runs in operations for 1 year (or 8760 hours).

Table 4-5 shows the TML solution costs.

Table 4-5. *TML Solution Costs[20]*

Cost Type	Amount (USD)	Throughput/Hour	Yearly Cost (USD)
Reads	$0.11/GB	1000	963,000
Writes	$0.11/GB	1000	963,000
Storage	$0.00013889/GB-hour	10,000	12,167
Partitions	$0.004/partition-hour	100,000	3,504,000
Total Annual Cost			$5,442,167

As shown in Table 4-5, for a large TML solution, the estimated run rate (for 8760 hours) is $5,442,167. This does not include the human cost as well as initial cost to develop the application. The main component of the total annual cost, about 65%, is the cost for Kafka partitions. Therefore, choosing the *right* number of Kafka partitions for each topic should be an important consideration when designing and architecting TML solutions.

While $5.5 million may not seem large for multinational organizations, this could further increase with the human costs of maintenance and support and additional costs for creating other TML solutions. Therefore, controlling ongoing costs by tracking TML solutions will become a key component in ensuring that the net benefits remain positive

---

[20]While these costs are notional costs for a large TML solution, they capture actual unit costs for the key components of a TML solution and will vary per solution.

over the life of the solution. Specifically, if the solution is not providing value or benefits to the organization that meets or exceeds $5.5 million, then organizations should be asking themselves "Can we improve the return on investment by scaling back the solution or improving its application?" If the answer is No, then ask "Can we really afford this solution?"

Using the AiMS Dashboard is one way to control costs. Recall that AiMS displays metadata on TML solutions. There are four key fields that will be important for tracking TML solution use, specifically:

- Bytes written (Kb)

- Bytes read (Kb)

- Last read of topic

- Last write to topic

TML solutions do not incur (cloud) costs if they are not running. But, in cases when TML solutions are running but no one is consuming from the solution, then you are wasting resources and cloud charges, shown in Table 4-4, will apply. To avoid this, TML solution administrators should set alerts to deactivate TML solutions that no one is using. These notifications can be configured to track the last time a topic was read from by a consumer. If no one has read from a topic (which is part of the TML solution) for, say, 30 days, then this topic should automatically be deactivated to stop it from consuming valuable cloud resources, saving your organization money. Therefore, tracking and monitoring TML solutions for business use or non-use, and then taking immediate action for non-use, could be an important cost controlling measure. The next section discusses the risks and pitfalls of TML solutions.

Risks and Pitfalls

TML solutions can be very powerful in helping to extract machine learning insights from data streams at scale. The downside of this power could be the cost of ownership if it is not managed properly. While cloud costs continue to come down, organizations must still pay close attention to the net benefits of TML solutions over time.

Organizations should be aware of key risks of TML solutions:

1) Reduction of net benefits over time – This can be due to TML solutions running, consuming valuable resources, but no one is using the solution. If no one is using the solution, then the solution is providing no value to the organization. Therefore, while TML solutions may provide initial value, over time this may decrease.

   Mitigation:

   a. Organizations can mitigate this risk by closely monitoring TML solution use using AiMS.

   b. Clearly understanding the value of the solution to the organization, over time, before developing the solution. This should be done by using the methods discussed earlier, such as mapping the business area, decision area, value area, and time to the business priorities in the BSC. Computing the TVI for each TML use case and then choosing the use case with the lowest TVI value will help to prioritize the highest value solutions. As well as checking in on the product in deployment by showing the number of users consuming from the solution, amount of data reads and writes, number of days the solution has been active in weekly reports. You could also determine if these numbers are aligning with expected value for the solution as stated in the original business case and if it is delivering value that was ascribed to it in the BSC. Using this information, you can make a decision to keep the solution running or deactivate it.

   c. Give access to TML solutions to users who need the outcomes for their business decision-making. Carefully controlling access to the solution can also help to reduce costs by optimizing the number of partitions needed in the topic.

2) Increase in the number of algorithms created – AutoML helps to reduce the friction from machine learning, but it also increases the number of algorithms in the organization. This could create challenges for

   a. Model management

   b. Model governance

   c. Security

   Mitigation:

   a) Better planning of the TML solution will be important. Specifically, understand how many algorithms will be needed for the solution, how many topics you will create in Kafka, how topics will interrelate, how much data will be read from topics and written to topics, and how many partitions will be needed for each topic.

   b) Use AiMS to deactivate topics that are no longer being used.

3) Stakeholder dissatisfaction with TML solution – TML solutions must be carefully thought out. This will involve choosing the right TML use case, number of topics to create, how topics will relate to each other, and number of partitions per topic.

   Mitigation:

   a) Choose TML use cases that have a low TVI score and aligned with the BSC.

   b) Identify who needs the outcomes of the solution for their business, and only give access to users who need it.

   c) Identify who will be producing to the topics. The producers could be humans or machines. This will help in supporting and maintaining TML solutions should something go wrong.

   d) Identify and compute the benefits and costs for the solution. Ensure benefits will outweigh the costs, *over time*.

4) Cost overruns – Not properly using the alerts/notifications in AiMS could increase the risk of "runaway" costs for TML solutions from topics/algorithms that no one is using and adding no value to the organization.

Mitigation:

a) Use AiMS to carefully track all topics.

b) Use alerts/notifications to keep up to date on last reads and writes of topics to ensure they are still being used.

c) Deactivate topics if not being used. Topics can always be activated again.

5) TML failure – Architecting TML solutions for load shedding in a distributed network can dramatically improve performance and usage. Otherwise, solutions will be slow, discouraging use.

Mitigation:

a) Understand how many people will use the solution and design it accordingly.

b) If the solution will run in a distributed network, then ensuring it is designed to load shed will help improve performance and increase usage. This may include multiple instances of VIPER and HPDE, together with several Kafka brokers or servers.[21]

c) If one topic will be accessed by multiple consumers, then create a consumer group to distribute information to consumers in parallel.

d) Choose the right amount of partitions for a topic if multiple consumers will use it. The number of partitions should equal the number of consumers for parallel processing.

e) Ensure redundancy is built into the system to prevent downtime in the event the server or network is down by using replication factor in Kafka.

---

[21]This will be discussed in Chapter 5.

Large TML solutions can get complicated quickly; however, taking a measured approach that aligns with organizational priorities will ensure wider acceptance of the solution. Avoiding the preceding risks by following the mitigation strategies will also aid in keeping costs down, and benefits high, over the life of the solution. The next section provides some concluding remarks.

Concluding Remarks

This chapter has discussed key value drivers of TML solutions, such as

1) Choosing the right *business areas*

2) Choosing the right *decision areas*

3) Choosing the right *value areas*

4) Choosing the right *time to deliver* the solution insights

It discussed how to align these drivers to corporate objectives and how to measure the value of a TML solution. The potential risks of any large technology solution are not only in the people and process but the technology itself. The value derived from the technology solution must be weighed against the cost of the technology over its life.

For TML solutions that rely on cloud resources such as storage and compute, costs of ownership must be monitored over time and continuously weighed against the value provided by the solution. In large organizations, it is easy for technology and solution costs to grow quickly, without equal value returning back to the organization to justify the ongoing operations of the solution. Monitoring TML solutions to determine whether the solution is being used will be important in determining whether the solution, and its component, should be deactivated to save money. This is prudent from a model management, and governance, perspectives.

Using the Algorithm and Insights Management System (AiMS) is one way to monitor and track TML solutions. Specifically, TML administrators can track (among others)

- Bytes written (Kb)

- Bytes read (Kb)

- Last read of topic

- Last write to topic

The AiMS fields can be used to determine the last reads and writes of topics that comprise a TML solution. Using the information on the last date and time of reads and writes, administrators can determine if anyone or any application is reading data or writing data to the topics. Administrators can set alerts to continuously check for reads and writes and can manually or automatically deactivate a topic that no one is using. Alternatively, if nothing is writing to the topic, this would also signify that something is wrong with the TML solution, which could be quickly rectified by people who developed the TML solution.

The risks and costs of TML solutions are muted for small TML solutions that may not need cloud resources. Large, cloud-based, TML solutions present the highest risks of runaway costs. If TML solutions are not effectively tracked and monitored continuously, they can incur excessive costs without the offsetting value back to the organization.

Up to this point, we have been discussing TML from a largely theoretical perspective and establishing its importance for businesses that want to make fast decisions from fast data. We have described the relationships between data streams and AutoML and how they comprise a TML solution. The next chapters will take a more technical look at TML. Specifically, we will discuss its architecture and components and show how we apply these components to build TML solutions with low code.

CHAPTER 5

The Technical Components and Architecture for Transactional Machine Learning Solutions

This chapter will move the conversation from theory to practice, to show you how to bring TML solutions to life, to present the architecture of a typical TML solution that would allow you to determine whether a TML solution is a good fit for the problems you are trying to solve, and also to show you how TML can integrate with other technologies, specifically how output from a TML solution could be used as input into other downstream systems. We will discuss the data flows between TML components and how these data are processed in each component. The next section provides an overview of a TML solution.

Overview of a TML Solution

The first step in building a TML solution is to identify whether TML is the right solution approach. As discussed in the previous chapter, there are four areas to consider when deciding upon a TML use case:

- Business area need for fast decisions
- Fast decisions to be made

© Sebastian Maurice 2021

S. Maurice, *Transactional Machine Learning with Data Streams and AutoML*,
https://doi.org/10.1007/978-1-4842-7023-3_5

- Value required to help your business

- Speed in realizing value from the decisions

A quick rule of thumb to determine if TML is the right solution approach is *if you have fast data, then you will need fast (transactional) machine learning. Fast data has three components: (1) speed of data accumulation, (2) speed of data flow between the source and sink, (3) speed of change in the variety of data (text or numeric).* However, while this is a rule of thumb, further consideration should be given to whether these data streams can be joined to create training datasets for machine learning. The training dataset will be required for the application of machine learning to find an optimal algorithm that is used to make predictions, find optimal values, and anomaly detection.[1] The search for an optimal algorithm, with fast data or data streams, is done using auto machine learning. Each TML solution will have the following components:

- Data stream producers

- Cloud or field gateways that act as pass-throughs for the data

- Middleware software that receives/sends the data for further processing in DSSP

- AutoML to perform machine learning on the training dataset

- AutoML to generate predictions, optimization, and anomalies for consumers

- Visualization of streamed results in real time

- Management of algorithms and insights

Specifically, data producers are continuously generating data. A data producer can be a human or a machine. The raw data is received by the middleware software (MWS) that produces it in the DSSP. The data are formatted in JSON by the MWS. The MWS also consumes data from the DSSP for processing. The processing converts the raw data (stored as JSON in the data stream) to a training dataset by joining data streams.

---

[1]Anomaly detection is performed using unsupervised learning by HPDE.

The AutoML technology consumes the training dataset[2] and applies machine learning to find the best algorithm that fits the data. The best algorithm is stored in the DSSP. The AutoML technology also consumes the optimal algorithm to generate predictions, optimal values, and anomalies which are then visualized and consumed by consumers. The data are stored in the cloud on AWS, Google, or Microsoft servers running Apache Kafka.

The consumers visualize the streamed results in real time. Specifically, the results are pushed to your web browser over a secure WebSocket via HTTPS, and tables and graphs are dynamically generated by the visualization technology. The advantage of WebSockets is that you do not need to request results from the MWS; the MWS will automatically push results to you as they become available in the DSSP. Administrators of TML solutions manage the algorithms and insights. They can control the operations of all components of TML solutions using AiMS. For example, if you are seeing a large amount of data reads for an anomaly solution (topic) and the last read of the topic from a consumer is days in the past, then you would want to investigate why the consumer is not consuming the latest results. If the consumer no longer uses the results, or left the company, this solution should be a candidate for deactivation. You could also automate these cases by configuring notifications in AiMS, and it will automatically monitor TML topics for last use and email you the results. The next section will discuss each of these steps/components in detail.

Reference Architecture of a TML Solution

A typical reference architecture of a TML solution is shown in Figure 5-1. The key components are the producers and consumers of data. This is a pub-sub model that is used by Apache Kafka. The data flows into the DSSP, and out from the DSSP, are managed by the MWS, which produces and consumes data from the DSSP.

---

[2]MAADS-VIPER will extract the data from the source, then transform JSON data to data arrays to form a training dataset, and then load the transformed data to the DSSP for TML. So, this will be an extract-transform-load process.

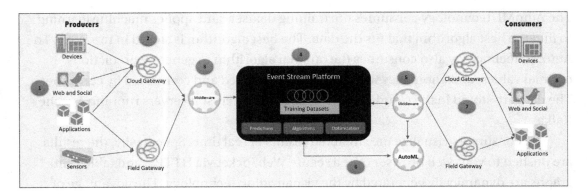

Figure 5-1. *TML Reference Architecture*

Table 5-1 describes each of the steps in Figure 5-1.

Table 5-1. *TML Process Steps*

Step	Description
1	These are data generators that produce data. The producers can be • IoT sensors • Software applications • Web and social media • Hardware devices: phones, watches, cars, TV, refrigerators, etc.
2	The data passes through a cloud or field gateway. Cloud gateways are routers, switches, etc. Field gateways can be MQTT, Modbus, and REST. These gateways have specific communication protocols that can convert electrical signals to digital signals required for processing. Internet protocols like HTTP, accessed via REST API, can handle cloud traffic.
3	Middleware software (MWS) is a specially designed software that connects to the DSSP to perform reads, writes, data transformations, makes calls to AutoML, all without human intervention. It acts as a source.
4	DSSP stores and manages all streaming data, algorithms, predictions, optimal values, anomalies, and training datasets. It is the central repository of all information. It also provides redundancy for failovers and scales as the data streams grow.

(*continued*)

Table 5-1. (*continued*)

Step	Description
5	Middleware software (same as in 3), which reads the training dataset from the DSSP and calls AutoML to initiate processing. It acts as a sink.
6	The AutoML, once initiated by the MWS, connects to DSSP to read the training dataset for processing. MWS passes the location of the training dataset in the DSSP to the AutoML technology; there is no training dataset exchange between the MWS and AutoML. Once the AutoML determines the optimal algorithm, it generates a unique key to identify the algorithm and stores this key in another topic in the DSSP for future use.
7	The MWS reads the optimal algorithm unique key and uses it to provide predictions and optimal values that are stored in a topic. Consumers subscribe to this topic and consume the insights through streaming visualization.
8	The predictions, anomalies, and optimal values can be delivered to devices and applications that adjust sensor readings or write to web and social media channels or used by other downstream applications.

Steps 4, 5, and 6 involve continuously reading and writing data to the DSSP as new data streams in. Specifically, recall that data streams are a continuous flow of data, so any machine learning process must also be continuous for the following reasons:

- Newer data may contain newer information, trends, and patterns that will require relearning.

- With relearning, the optimal algorithms will capture different information and may provide different predictions, optimal and anomaly values, or peer groups.

- Newer predictions and/or optimal values or anomalies will influence decision-making.

The preceding reasons are also why TML differs from CML: quickly learning from streaming data can identify new and emerging patterns in the data, which would not be possible using slower CML processes. The speed of learning from data streams is a fundamental advantage of TML over CML. While CML can retrain or relearn from new data, it will need to first extract the historical data, combine it in a training dataset, perform any transformations, apply machine learning algorithms to the training dataset, fine-tune the hyperparameters, determine the optimal algorithm, then deploy the optimal algorithm, and finally start using it for predictions, optimization, or anomaly detection. All of this will take a lot of time: days or weeks. In the days or weeks it takes to deploy a new retrained model, the underlying structure in the data may have already changed, making the CML solution less useful and effective. But, while CML is a longer process, it could be used to optimize the TML model. For example, in anomaly detection, you can set parameter values that determine what is anomalous or not; you can adjust these parameters to be very sensitive or less sensitive. The CML model could be used to determine parameter values that provide the required amount of sensitivity in the TML model.

There are other benefits from the TML process as outlined in Table 5-1. Specifically

- Model management – As AutoML retrains on newer data, it may be that the old algorithm is no longer the "best" algorithm, and a different algorithm is *better*. This different algorithm is stored in the DSSP, along with the older one. This simplifies model management because the newer model can be accessed exactly the same way as the older model. The older model can still be referenced and used to compare and contrast results with the new model.

- Audit trail – As newer algorithms, or hyperparameters, are found from retraining on newer data, they are appended to the same topic or another topic; this ensures older models are not overwritten or discarded. This allows organizations to maintain an audit trail of all the algorithms that have been found on training datasets that originated from the same data streams. This could be important if a wrong decision is made from a TML solution using this algorithm that results in a financial loss or some other incident. Having the ability to trace back to the root causes of failures will help to fix, improve, and recalibrate the TML solutions for better future decision outcomes.

- Compare model results – The other benefit of storing algorithms in the DSSP is for model comparison. The ease of access of algorithms allows users to see how different models perform with similar data. This can also be used to test outputs from models that others have built with the AutoML technology outputs.

- Combine model outputs – Combining the outputs of different models into one consolidated model could allow users to see how the overall system performs. Specifically, TML solutions can be built to solve smaller problems that when combined form a larger system. This system of systems approach with TML could be very effective in breaking down very large machine learning problems into subcomponents for more efficient analysis. For instance, for a large fraud campaign, multiple TML solutions could be analyzing distributed transactions in each jurisdiction (countries or cities) to predict anomaly risk scores that are stored in their respective data stream, that is, Canada data stream, US data stream, UK data stream, India data stream, and so on. These data streams could then be combined by another TML solution to represent the global (overall) risk score.

- Construct training datasets – Outputs of one TML solution can be an input into another. In this way, solution outcomes can be combined to construct a specific training dataset by deconstructing a system into submodels. The level of information from these submodels could contribute significant information to the overall model or system. For example, if I am predicting electricity prices for tomorrow, I can build TML solutions to predict weather temperature and electricity demand, using their outputs as input into my third TML solution to predict tomorrow's electricity price.

The next section will discuss the technical components of a TML solution.

Description of Technical Components

There are several components that are required to build a TML solution. This section will describe each one in detail, with the names of the technology. The commonality between every TML solution is

- Data streams are joined to construct training datasets.

- Offsets are used to roll back the data stream to get historical data. Recall that offsets are simply numbers that indicate the position of the data (consumer) in the data streams, that is, offset=0 for a data point (consumer) means it is the first data point in the data stream. Offsets increment sequentially in each partition in the data stream.

- Supervised learning

    a. AutoML technology applies machine learning on the training dataset to produce an optimal algorithm.

    b. AutoML uses the optimal algorithm to make predictions or find optimal values.

- Unsupervised learning

    a. AutoML technology applies machine learning to find peer groups for anomaly detection.

    b. AutoML determines the likelihood of an anomaly in new data by comparing it with the peer groups in real time.

- Scale and repeat the above with more data streams, algorithms, and peer groups.

Table 5-2 describes each technical component, its prerequisites, and core functionality.

Table 5-2. *Technical Component Details*

Technical Component	Description	Prerequisites	Function	Protocol
DSSP	Apache Kafka (on-premise or cloud)	**On-premise:** • Computer server with Windows or Linux operating system • Internet is not required **Cloud:** • Confluent Cloud running Apache Kafka • Amazon AWS running Kafka • Microsoft Azure running Kafka • Google Cloud running Kafka • Internet connection for cloud	• Manages offsets • Manages partitions • Manages replication of data for failover • Stores all data streams • Stores all training datasets • Stores all algorithms • Stores all predictions • Stores all optimal values	• TCP/IP • SSL/TLS (optional)

(continued)

117

Table 5.2. (*continued*)

Technical Component	Description	Prerequisites	Function	Protocol
MAADS-VIPER	Apache Kafka connector: source and sink	**On-premise:** • Computer server with Windows, MacOS, or Linux operating system **Cloud:** • Computer running Windows, MacOS, or Linux OS • Apache Kafka	• Creates topics in DSSP • Reads from DSSP • Writes to DSSP • Creates training datasets • Subscribes consumers to topics • Creates producers • Creates consumer groups • Initiates AutoML • Performs predictions • Performs optimization • Integrates with AiMS • Performs AiMS functions: activate or deactivate topics • Alerts and notifications	• TCP/IP • HTTP • SSL/TLS (optional)

Component	Description	Requirements	Functions	Protocols
MAADS-HPDE	AutoML for TML, performs supervised and unsupervised learning	**On-premise:** • Computer server with Windows, MacOS, or Linux operating system **Cloud:** • Computer running Windows, MacOS, or Linux OS • MAADS-VIPER	• Connects to DSSP to read training dataset • Performs AutoML on training dataset • Finds optimal algorithm that best fits the data • Writes optimal algorithm to DSSP • Stores optimal algorithm to disk • Reads optimal algorithm from disk • Performs predictions • Performs optimization • Performs anomaly detection	• TCP/IP • HTTP • SSL/TLS (optional)
AiMS	Manages algorithms stored in DSSP	• MAADS-VIPERviz • Windows, MacOS, or Linux OS	• Manages algorithms generated by HPDE • Activates or deactivates algorithms • Performs alerts and notifications • Creates incidents in service management platform • Manages Kafka consumers, producers, and groups	• HTTP • HTTPS • TCP/IP
MAADS-VIPERviz	Provides streaming visualization of supervised and unsupervised TML solutions	• Windows, MacOS, or Linux OS • Apache Kafka • MAADS-VIPER	• Provides all visualization from HPDE • Uses WebSockets to push new results to client web browsers	• HTTP • HTTPS • TCP/IP

(continued)

119

Table 5.2. (*continued*)

Technical Component	Description	Prerequisites	Function	Protocol
MAADS Python library	API for TML solutions	• Computer running Python programming language for Windows, MacOS, or Linux interpreter • Jupyter Notebook (optional) • MAADS-VIPER	• TML functions for VIPER	• HTTP • TCP/IP (optional)
REST	REST API	• Any computer running Windows, MacOS, or Linux • Any programming language that supports REST (i.e., HTTP requests) • MAADS-VIPER	• TML functions • REST points to a server running VIPER on a HOST and PORT	• TCP/IP • HTTP

The components discussed in Table 5-2 can be instantiated for scale. Specifically, for large TML solutions, it may be required to load shed by having several instances of DSSP, VIPER, VIPERviz, and HPDE. Each of these applications can scale to any number of instances only limited by hardware. The next section will show how these components fit together within a cloud instance.

Technical Architecture of a TML Solution

The previous section discussed the technical components that make up a TML solution. In this section, we show how these components fit together in a technical architecture shown in Figure 5-2. The first thing to note in Figure 5-2 are the swimlanes.

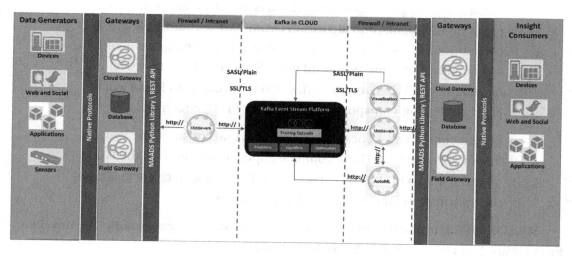

Figure 5-2. *TML Technical Architecture*

- Data generation – These are all the applications, legacy devices, mobile devices, sensors, and the Web and social media.

- Gateways – All data flow through either cloud gateways or some field gateways. These gateways connect the core networks to edge networks and provide access controls that allow data to move through the network to the end users. These also connect different cloud/data centers in different locations. Or, these can store data in a database that can be accessed by the TML solution. TML will perform the following data transformations in real time: (1) joining multiple data streams, (2) pruning each data stream to equal lengths to form the training dataset, (3) converting the JSON objects in the data streams to data arrays for machine learning, (4) constructing the peer group using Z-scores and peer group analysis for anomaly detection, (5) consolidating and formatting new data streams for predictions, optimization, and anomaly detection.

- Firewalls – Ensuring data entering and leaving a data center (or network infrastructure) is secure with appropriate authorization is moderated and controlled by a software and hardware firewall. Organizations that want to secure their environment will establish a firewall perimeter that controls data entry and exit.

- Kafka in the cloud – Kafka in the cloud is our DSSP and will store all of our data, insights, and algorithms from TML solutions for end user consumption. The Kafka Cloud is secure, and accessing it will require a cloud username and password using SASL/PLAIN authentication. As well, all traffic between a TML solution to and from Kafka will be encrypted using Transport Layer Security[3] (TLS) encryption.

- Insight consumers – These are consumers that consume the insights from TML solutions. These could be applications, devices, the Web, or social media users. You can be an end user or a machine that takes the information for decision-making purposes.

MAADS-VIPER, inside the firewall, receives data from data generators. The raw data format must be in text format. MAADS-VIPER will then

1) Use the MAADSTML Python library function **vipercreatetopic** to

   a. Log in to the Kafka Cloud using a cloud username/password

   b. Create a topic in Kafka

2) Use the MAADS Python library function **viperproducetotopic** to

   a. Log in to the Kafka Cloud using a cloud username/password

   b. Convert the data to JSON

---

[3]www.rfc-editor.org/info/rfc8446

    c. Encrypt the data using TLS

    d. Write the JSON data to the Kafka topic created

To process data, in the Kafka topic, for supervised learning, MAADS-VIPER will read the topic(s) in Kafka following similar login procedures and

3) Use the MAADS Python library functions

    a. **Vipercreatejointopicstreams** – To join the multiple streams of data and create a topic for the joined streams

    b. **Viperproducetotopicstream**

       • Consolidate the stream data

       • Roll back each stream based on user selection of offset to form the historical data

       • Produce data from all streams to the joined topic stream

    c. **Vipercreatetrainingdata** – Create the training dataset and produce it to another topic

4) Once the training dataset is built and saved in another topic, it can be processed by calling HPDE. Specifically, using the function **viperhpdetraining**:

    a. Log in to Kafka Cloud to retrieve the training dataset (there is no training dataset movement from VIPER to HPDE) and apply machine learning algorithms to the dataset.

    b. It will generate an optimal algorithm, which it physically writes to disk, and will store all metadata about the algorithm in another Kafka topic.

    c. It passes back to VIPER the name of this Kafka topic, which VIPER can use for predictions and optimization.

5) Making predictions with optimal algorithms is done using **viperhpdepredict**:

    a. This function will accept input data, that is, the values of the independent variables, or another data stream, pass the input data to the optimal algorithm, and return a prediction.

b. This function can be put inside a loop to continuously pass input data to the optimal algorithm to generate the prediction. Or, users can point the input data to a data stream and generate continuous predictions as new data enters the input data stream.

6) If users need to find the best or worst values of the independent variables, they use **viperhpdeoptimize** to perform mathematical optimization to find the optimal values for the independent variables.

   a. The optimal values are stored in another Kafka topic and retrieved by applications to use; these values (and approach) would be ideal for building closed-loop solutions where a machine makes all the decisions without human intervention.

7) Consumers, via MAADS-VIPER, consume the predictions and optimal values using **viperconsumefromtopic**.

   a. This function logs in to Kafka Cloud and uses the OFFSET parameter to read data starting from the OFFSET number. Users can read data in topics from the beginning (OFFSET = 0) or from the end (OFFSET = -1) or anywhere in between.

   b. The data are in JSON format and can be read by almost any programming language or application.

   c. The prediction value or optimal values can be easily extracted from the JSON data.

8) For cases when there are a very large number of consumers that want to access the *same* topic, then **viperconsumergroupconsumefromtopic** can be used.

   a. This function can be used for parallel processing of data in a topic. As data is written to a topic, this function will distribute the data to each consumer *at the same time*.

   b. To optimize a topic for parallel processing, the topic should have the same number of partitions as the number of consumers. This allows Kafka to distribute data to each partition, in the topic, so that it can be accessed simultaneously by each consumer for consumption. The administrator can create a topic with an initial number of partitions and, if needed, increase this number using the MAADSTML Python library if more partitions are needed to accommodate more consumers.

c. This function can be very useful to improve performance of large, distributed, TML solutions.

Unsupervised Learning

To process data using unsupervised learning for anomaly detection, you would use two functions: **viperanomalytrain** and **viperanomalypredict**. The process is similar to that of supervised learning, but here we do not create a training dataset. Rather, we use viperanomalytrain to find the peer groups from the data streams that we want to analyze for anomalies. Specifically, the process to generate peer groups from data streams is as follows:

1. **vipercreatejointopicstreams** – To join the multiple streams of data and create a topic for the joined streams.

2. **viperproducetotopicstream**

   a. Consolidate the stream data

   b. Roll back each stream based on user selection of offset to form the historical data

   c. Produce data from all streams to the joined topic stream

3. **viperanomalytrain** – Creates a peer group of all the streams. This peer group is a group of transactions that represent "normal" behaviors.

4. **viperanomalypredict** – Generates the probability scores for each new data that is tested for anomalies by comparing it to the peer group – in real time.

We will show how the preceding process comes to life in the next chapter. Before we get there, the next section discusses the communication process between TML components.

Communication Process Between Components

Effective communication between TML components is critical for a successful TML solution. There are three important components to ensure effective communication:

- Security

- Concurrency

- Algorithm management

All communication between Kafka, VIPER, VIPERviz, and HPDE are SSL/TLS encrypted. For sensitive information traveling across a distributed network, this security will be important. Concurrency is also an important aspect to ensure TML solutions can handle multiple producers and consumers that want to write and read information from Kafka, simultaneously. While multiple instances of VIPER, HPDE, and VIPERviz can be stood up, each application has concurrency built in to handle multiple requests at the same time. This not only improves performance but enables fewer instances of each application to reduce solution overhead and costs. Figure 5-3 shows the communication cycle for a TML solution.

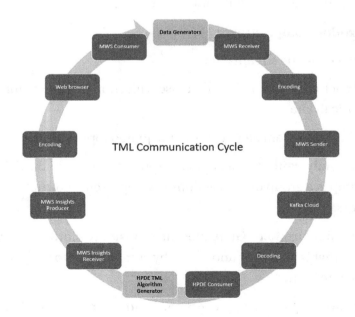

Figure 5-3. *TML Communication Cycle*

Figure 5-3 shows 12 steps in the TML communication cycle. These are described as follows:

- **Data Generators** starts the communication cycle by producing data to Kafka.

- **MWS Receiver** receives data from data generators.

- **Encoding** MWS uses SSL/TLS encryption to encode the data and sends it to a Kafka topic.

- **MWS Sender** sends data to a Kafka topic.

- **Kafka Cloud** stores the data.

- **Decoding** MWS decodes the location of the training dataset in Kafka and sends it to HPDE. Note: MWS does not send the training dataset to HPDE, just its location in Kafka.

- **HPDE Consumer** consumes the data in Kafka and performs machine learning on the training dataset.

- **HPDE TML Algorithm Generator** generates an optimal algorithm, fine-tunes the hyperparameters, and writes it to a Kafka topic.

- **MWS Insights Receiver** receives the predictions, optimizations, and anomalies in Kafka.

- **MWS Insights Producer** produces the predictions, optimizations, and anomalies in real-time graphs and tables that are pushed to the users' browser using WebSockets.

- **Encoding** MWS communicates with the client's web browser using WebSockets to push insights over HTTPS for visualization in the client's browser.

- **Web browser** displays real-time predictions, optimizations, anomalies, and AiMS.

- **MWS Consumer** consumes the visualizations on predictions, optimizations, anomalies, and AiMS from their web browser.

The cycle can be scaled to any number of data streams (data generators), algorithms, producers, and insight consumers. The Kafka Cloud platform allows for horizontal scalability making TML very effective in handling large amounts of fast data that require fast learning.

The next section will discuss data flows between the TML solution components. It will show how, and what types of, data are produced and consumed. It will also show the types of data that need to be produced and consumed by consumers to be used for visualization and decision-making.

Data Flows

Data flows between TML components are standardized to JSON format. The JSON format is a widely used and accepted data standard for storing, displaying, and analyzing information. Kafka is also compliant with JSON standards. Additional advantage of JSON data is cross-platform and cross-application compliance, meaning if results from a TML solution need to be used by another system, in many cases, these systems should be able to read JSON formatted data. Another important aspect of JSON is performance. Parsing data in JSON format is much more efficient than non-JSON data. As well, storage of metadata information on every algorithm created by HPDE is stored in JSON, in an embedded database, read by VIPER. This makes portability of VIPER and HPDE data across different operating systems more seamless and efficient.

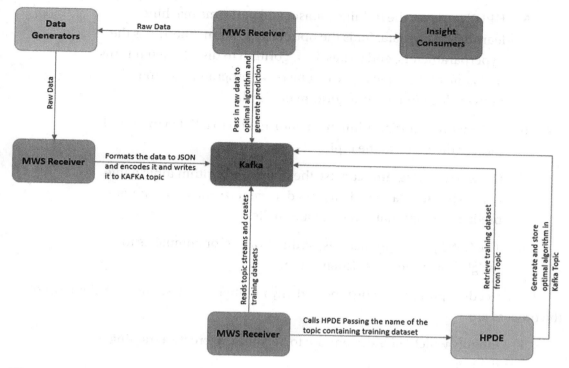

Figure 5-4. *How Data Flows Between TML Components*

Figure 5-4 shows the movement of data between the core TML components, specifically:

1. Data generators send raw data received by the MWS (MAADS-VIPER).

2. MWS formats the data in JSON and records the date, time, location, value, and other information in key-value pairs, encodes the data, and produces it to a Kafka topic. Note there can be multiple topic streams being written (potentially at the same time) to Kafka.

3. MWS receiver consumes the topic streams and creates training datasets using the MAADS functions described earlier.

4. MWS makes a call to HPDE that is listening for connections on a host and port. MWS passes HPDE the name of the topic in Kafka containing the training dataset.

5. HPDE retrieves the training dataset and performs machine learning on the data, finds the optimal algorithm, fine-tunes the hyperparameters, and writes the algorithm to disk. It then returns the name of the topic to the calling MWS program, with the physical location of the algorithm on the disk.

6. MWS writes the algorithm meta information to a Kafka topic and returns the name of the topic.

7. Other MWS programs can use the optimal algorithm topic name to pass input data (raw data from data generators) and generate predictions, optimal values, or anomalies.

8. The MWS passes predictions, optimal values, or anomalies to insight consumers for visualization.

The preceding process can be repeated any number of times. The core elements of the process are

- Identify which data streams are to be joined to create a training dataset

- Record the optimal algorithm names (i.e., topics in Kafka) for each of the training datasets

- Produce to and consume from the right topics

- Use the new raw data with the right optimal algorithm to produce predictions and/or optimal values and anomalies

- Send the right predictions and/or optimal values or anomalies to the right consumer applications, devices, or the Web/social media for consumption

As a consequence, to the preceding process, TML solutions will

1. Create several topics.

2. Generate several optimal algorithms written to disk.

3. Topics may require several Kafka partitions.

4.   Use Kafka storage and throughput.

5.   Incur cloud costs.[4]

Therefore, administrators, solution architects, and visual designers should pay careful attention to the overall solution costs and minimize any unneeded topics, partitions, and storage. The next section discusses and shows an example TML solution architecture.

Example Architecture

Figure 5-5 shows an example architecture using Confluent Cloud.[5] Confluent Cloud is a platform that provides Kafka on Amazon AWS, Microsoft Azure, and Google Cloud as a managed service.

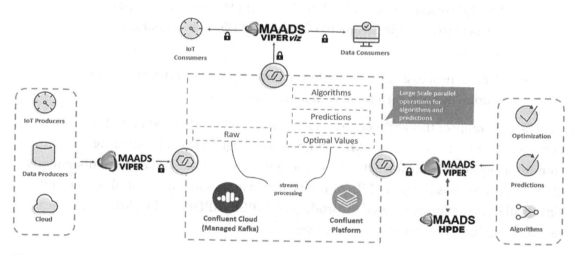

Figure 5-5. *TML Example Architecture*[6]

---

[4]See Table 4-4 TML Cost Components and Table 4-5 TML Solution Components.

[5]See www.confluent.io/confluent-cloud.

[6]See www.confluent.io/blog/transactional-machine-learning-with-maads-viper-and-apache-kafka/?utm_source=linkedin&utm_medium=organicsocial&utm_campaign=tm. devx_ch.transactional-machine-learning-with-maads-viper-and-apache-kafka_content. analytics- for more details.

Some notable features about the preceding architecture:

- The training dataset is stored in Kafka running in the Confluent Cloud using the Amazon, Microsoft, or Google infrastructure.

- MAADS-VIPER (MWS), VIPERviz, or HPDE run in Windows, Linux, or MacOS environments.

- MAADS-VIPERviz is visualization technology for visualizing streaming results in Kafka.

- All optimal algorithms, predictions, anomalies, and optimal values are stored in Kafka.

- SSL/TLS encryption is used to secure data to and from Confluent Cloud.

- Multiple instances of VIPER, VIPERviz, and HPDE can be used to connect to multiple instances of Confluent Cloud in a distributed network.

- Parallel processing for multiple consumers accessing the same topic is done by creating a topic with multiple partitions.

The generality of this architecture, across any cloud vendor that provides a managed Kafka service, makes TML solutions cloud agnostic while ensuring standard communication and data format across different managed Kafka services. This extends the viability and feasibility of TML solutions. It also makes Kafka a central source and sink of all information generated by producers, VIPER, HPDE, and VIPERviz. The next section describes the TML cost components.

TML Cost Management

We discussed the TML cost components in the previous chapter; we expand on this here. Managing the costs of TML components, for cloud-based solutions, will be dependent on several components discussed as follows:

- People – Because TML automates many conventional data scientist tasks, you should plan for the following resources for designing, building, testing, deploying, and maintaining TML solutions. The number of people will depend on the size and quantity of TML solutions you plan to develop. The roles are the following:

  - Data stream engineer – A person who sources the data for the TML solutions and produces it to Kafka using MAADS-VIPER and MAADSTML Python library. This person should have knowledge of Apache Kafka and Python.

  - TML solution architect – This is a critical role because this person will architect how the TML solution will use the data streams, specifically what the source and sink streams are and how they will be joined to produce the outcomes for visualization, scale, and decision-making.

  - Data stream scientist – A person who builds out the approach, methodology, and model formulations of the TML solution and develops it using the MAADSTML Python library for deployment.

  - Data stream visualization expert – A person who maintains or enhances the out-of-the-box visualizations. MAADS-VIPERviz comes with four prebuilt visualizations for predictions, optimization, anomalies, and generic visualization that allow you to visualize combined data streams.

  - TML administrator – An operational person who manages TML solutions using the AiMS Dashboard.

  - Cloud infrastructure expert – A person who builds and manages the cloud infrastructure for the TML solutions along with defining the security and access control lists.

- Process and change management

  - Change management expert – A person who communicates the purpose and value of TML solutions to (business) users and defines any changes in the process required by TML solutions.

- Technology

  - Kafka partitions – The most important cost component is the number of partitions that you are using per topic. The cost is computed on a per partition hour basis. For example, 1000 partitions, for 1 topic, for 1 hour will cost you $4. While this may seem low, if you have the TML solution continuously running for 6 months with 1000 partitions, you will incur a $17,280 bill in 6 months for just 1 topic. So, use partitions carefully.

  - Network reads and writes – This will depend on how fast you want to read data and write data to Kafka. The higher the speed, the more the cost.

  - Storage – While storage is the lowest cost component, it should be managed accordingly.

The preceding components the core components to develop, deploy, and maintain TML solutions in an organization. There may be more components, but the preceding components represent the majority of the consideration that you need to make for a successful TML deployment. The next section concludes this chapter.

Concluding Remarks

This chapter has provided details on the TML reference and technical architectures, highlighting the communication and data flows between TML components. It also showed an example of the TML architecture. Some of the main areas of focus discussed are

- Communication cycle between the TML components such as data generators, VIPER, VIPERviz, HPDE, Kafka, and insight consumers. This can be a continuous process dependent on how the TML solution is developed using the MAADSTML Python library functions (or REST API). These functions provide developers access to VIPER and control its capabilities: producing to Kafka, consuming from Kafka, and calling HPDE. Powerful TML solutions can be developed with these library functions. Several TML solutions can be created and joined. Specifically, using the topics in Kafka, other TML solutions can use the same optimal algorithms and insights and share the information across different applications, devices, and the Web and social media.

- Data flowing between TML components are in the JSON format. The JSON format makes it easy to share insights across different applications and simplifies information retrieval from JSON objects. The standardization of data also makes it easy to manipulate the JSON objects such as combining and analyzing different objects. It also removes the requirement of maintaining a physical database schema for TML solutions, as metadata information is stored in an embedded database for VIPER and HPDE, making TML solutions highly portable across different environments.

TML solutions can be built as closed-loop or open-loop solutions. For closed-loop solutions, this allows TML solutions to be completely frictionless, and automated, removing any human intervention. The insights generated can be used to adjust downstream systems automatically. Open-loop solutions will require some human intervention in the decision-making process, but this is minimal compared to CML processes.

As architects design TML solutions, they should pay attention to the following:

- Cost of TML solutions is mainly dependent on

    a. People – Such as data stream engineer, TML solution architect, data stream scientist, data stream visualization expert, TML administrator, cloud infrastructure expert, and process and change management expert

    b. Kafka partitions – The higher the partitions, the higher the costs

     c.   Amount of storage

     d.   Number of topics

     e.   Data throughput across the network

- Parallel processing should be considered if multiple consumers access the same topic. This will require that each topic have several partitions to match the number of consumers. However, the higher the partitions, the higher the cost for the TML solution.

- The number of algorithms created may rise and complicate model management and may create security issues. TML administrators should use AiMS to track and manage algorithms and deactivate algorithms if they are not being used.

TML solutions are cloud agnostic, programming language agnostic, database agnostic, and operating system agnostic. They are compatible with any managed Kafka service on Amazon AWS, Microsoft Azure, and Google Cloud. Using the MAADSTML Python library, developers can create powerful TML solutions, with low code, that can scale quickly with unlimited amounts of data streams, consumers, and producers. The flexibility of the JSON data format makes it easy for diverse applications and devices to consume the information for machine or human decision-making and visualization. It also makes it easy to further manipulate JSON objects by combining different objects and analyzing those objects for further information extraction.

Scaling TML solutions by using multiple instances of Kafka, VIPER, VIPERviz, and HPDE makes it possible to build massive TML solutions. Even as business needs evolve over time, TML solutions can also grow (or shrink): this is the elastic nature of TML solutions. More data streams can be added, more training datasets can be created, and more algorithms can be generated. While this growth can add to costs and complexities inside an organization, the complexity and costs can be controlled, tracked, and managed using AiMS. The solution possibilities with TML solutions are almost endless, but together with AiMS they can be built cost effectively and responsibly.

CHAPTER 6

Transactional Machine Learning Solution Template with Streaming Visualization

This chapter will discuss in detail the technologies needed to build TML solutions with example code for

- Predicting and optimizing foot traffic in 11,000 Walmart locations
- Predicting anomalies in 50 bank accounts with 10,000 bank transactions

The intent of this chapter is to give you a hands-on walk-through of how to build your own TML solutions integrated with Apache Kafka with data streams and AutoML, using Confluent Cloud running on Google Cloud Platform. You will understand the core library functions and how they can be called with the MAADSTML Python library[1] that connects to VIPER and Kafka Cloud to perform very advanced auto machine learning for predictive and prescriptive analytics, as well as anomaly detection using unsupervised machine learning.

You will get familiar with common steps that are taken to build and optimize a TML solution. The solution template is a simple yet powerful way to help you build TML solutions quickly that can be customized for your own needs. You do not need to be an expert programmer; beginners will find it easy to build TML solutions with this

---

[1] https://pypi.org/project/maadstml/

© Sebastian Maurice 2021

S. Maurice, *Transactional Machine Learning with Data Streams and AutoML*, https://doi.org/10.1007/978-1-4842-7023-3_6

template. We will use Python as our programming language to access VIPER, but since VIPER can be accessed by REST calls, you can use any programming language you like as long as it supports REST, which is a majority of programming languages in use today. At the end of this chapter, you would have used the leading cloud platform, with leading Kafka technologies to perform advanced auto machine learning with (big) data streams, while using advanced visualization for streaming insights. All of the examples and technologies can be found on GitHub.[2] The next section provides an overview of the TML solution template and the accompanying technologies.

Overview of TML Solution Template

This template can be used to build any type of TML solutions that are scalable, frictionless, and elastic. As noted, all the following components described (with the exception of Kafka Cloud) can be downloaded from GitHub. The main template components are the following:

- Kafka Cloud – Needed to create a Kafka cluster (broker) running on Amazon, Microsoft, or Google infrastructure using Confluent Cloud with SSL/TLS security.

- MAADS-VIPER (just VIPER) – Official Kafka source and sink connector running in Linux (Windows and MacOS versions also available); it will perform all of the functions needed for TML solutions. This Kafka connector has been verified and tested by Confluent.

- Environment file – VIPER environmental file containing all necessary credentials including the token file (admin.tok) and example SSL/TLS security certificates: client.cer.pem, client.key.pem, server.cer.pem.

- MAADS-HPDE (just HPDE) – AutoML technology that performs TML running in Linux (Windows and MacOS versions also available).

---

[2]https://github.com/smaurice101/transactionalmachinelearning

- MAADS-VIPERviz (just VIPERviz) – Visualization technology to visualize: anomaly detection, predictions, and optimization results produced from HPDE. VIPERviz is also needed for AiMS.

- AiMS – Algorithm and Insights Management System dashboard, used to manage TML solutions.

- MAADSTML Python library, Python[3] IDE, and Jupyter Notebook.

As discussed, when building TML solutions, careful consideration should be given to the size of the TML solution and how it is being used over time. The size will be dependent on

- Number of data streams (topics)

- How many partitions in each data stream (topic)

- Number of training datasets for supervised learning or peer groups for unsupervised learning

- Number of algorithms, peer groups, predictions, optimal values, and anomalies generated

- How many consumers will consume from topics

- How many producers will produce to topics

- How many instances of Kafka brokers, VIPER, VIPERviz, and HPDE

- The size of the Kafka cluster

Each of the template components will be discussed in detail in the next section.

Template Component Details

This section will detail each of the template components and how it is used. It will provide you with all the necessary setup instructions needed to run the technologies required to build TML solutions. Each component provides advanced functionality that is built for fast data and fast machine learning. This may be new to many readers, but this

---

[3]https://pypi.org/project/maadstml/

chapter will do its best to make sure you understand each component and its purpose. After you understand the components, building TML solutions rely on a handful of functions that, when used properly, will allow you to build very large solutions in a very short amount of time with relative ease. Let's begin.

Kafka Cloud via Confluent Cloud

Kafka is an increasingly popular, and powerful, stream processing software platform.[4] Many major cloud vendors such as Amazon, Microsoft, and Google provide Kafka application services running in their infrastructure. This makes it very easy to build TML solutions that are **cloud agnostic**. VIPER is compatible with Kafka running on Amazon AWS, Microsoft Azure, Google Cloud Platform (GCP), or Confluent Cloud.

A very popular option is to use a Kafka managed service provided by Confluent called Confluent Cloud (CC). CC allows TML developers to choose Kafka instances to run on their choice of cloud vendors. Developers can choose to create Kafka clusters using Amazon, Microsoft, or Google Cloud infrastructure. The advantage of using CC to run Kafka on your cloud vendor of choice is threefold:

- Organizations that already use Amazon, Microsoft, or Google can continue to use them without experiencing any disruption to their business; as well, there isn't a need to learn a new technology.

- For the Enterprise setup of Kafka, you continue to get one cloud bill from Amazon, Microsoft, or Google – with Confluent as a line item in your bill. Specifically, each cloud provider will have marketplace connectors for Confluent Cloud that you would use to integrate your infrastructure billing with CC as a service provider. More details can be found on the Confluent website.[5]

- No special Kafka setup required – CC takes care of all the setup details, allowing developers to quickly focus on building their TML solution.

---

[4]www.businesswire.com/news/home/20180626005510/en/Survey-Reveals-Apache-Kafka%C2%AE-Will-Be-Mission-Critical-to-90-Percent-of-Data-and-Application-Infrastructures-in-2018

[5]https://docs.confluent.io/cloud/current/billing/overview.html

To start building your TML solution, you first need to create a Kafka cluster. To do this, follow these steps:

1) Go to Confluent Login[6] – If you don't have an account, please create one. If you already have an account, then log in.

2) Confluent will give you free cloud credits. This will allow you to build and test a reasonably sized TML solution and familiarize yourself with Confluent Cloud.

3) Once logged in, create a Kafka cluster, as shown in Figure 6-1.

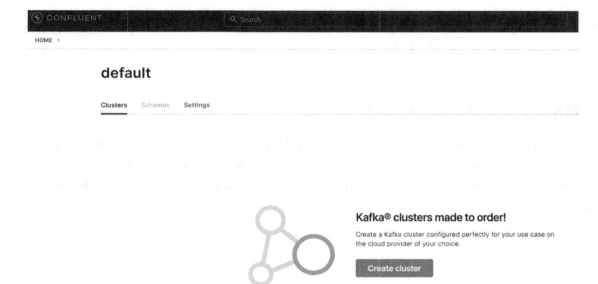

Figure 6-1. *Kafka Cluster*

Then, choose the Basic cluster settings as shown in Figure 6-2.

---

[6]https://confluent.cloud/login

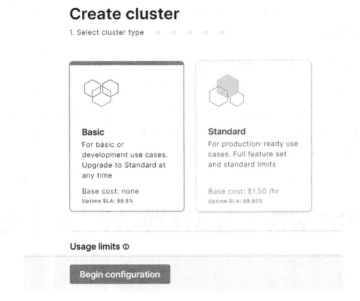

Figure 6-2. *Kafka Basic Cluster*

After you click Begin configuration, you will see the screen shown in Figure 6-3. This is where you choose between Amazon, Google, or Microsoft. You can choose anyone you like; however, if your organization is partial to a particular cloud vendor, then you should choose that one.

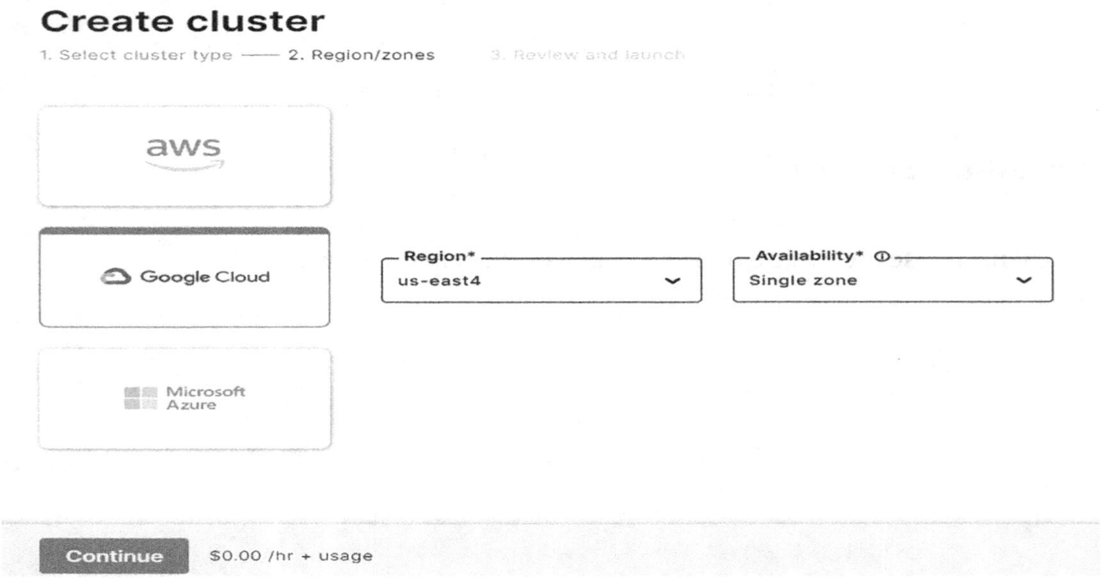

Create cluster

1. Select cluster type ——— 2. Region/zones   3. Review and launch

aws

Google Cloud

Region*
us-east4

Availability* ①
Single zone

Microsoft
Azure

Continue $0.00 /hr + usage

Figure 6-3. *Infrastructure Choice*

Once you have chosen a particular cloud vendor, then press continue. On successful cluster creation, you will see Figure 6-4.

Figure 6-4. *Cluster Overview*

There are a few additional steps after a cluster is created. To access the cluster from VIPER and HPDE, we need to create an API key. To do this, click **Create an API key associated with your account** as shown in Figure 6-5.

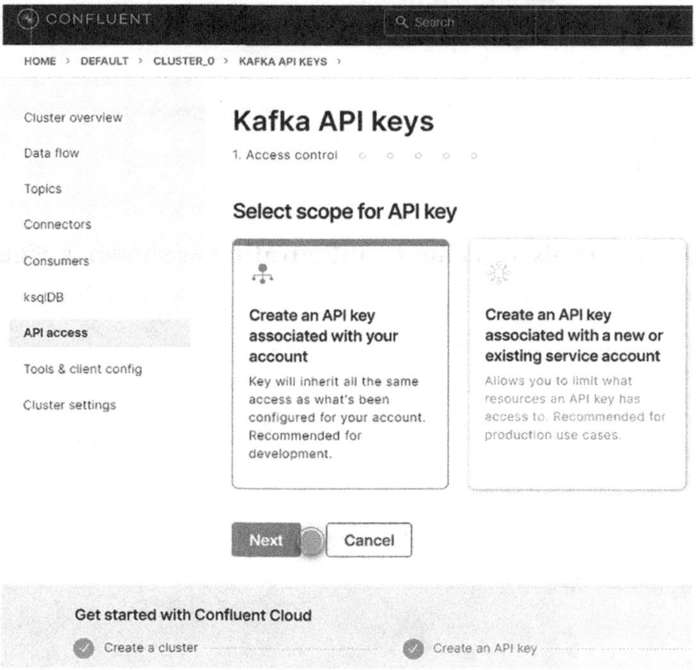

Figure 6-5. *API Key*

After clicking **Next**, you will generate two keys as shown in Figure 6-6:

- Key

- Secret

Make sure to record this Key and Secret. These will be your cloud credentials – I have grayed out my keys for security reasons.

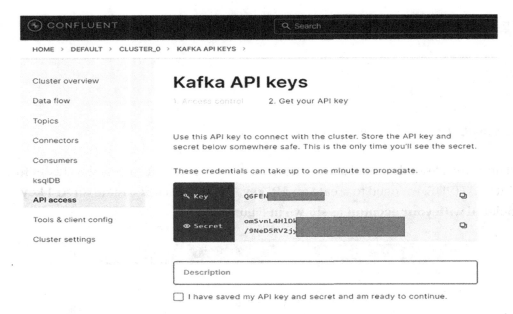

Figure 6-6. *Key and Secret*

The last step is to go to **Tools and client configuration** as shown in Figure 6-7.

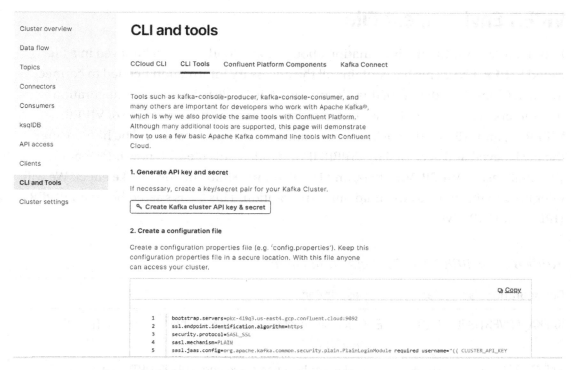

Figure 6-7. Tools and Client Configuration

In Figure 6-8, scroll down to **Go** and copy the ***bootstrap.servers*** address.

Figure 6-8. Bootstrap Server

Congratulations, you have successfully configured your Kafka cluster in the cloud. Next, we need to set up MAADS-VIPER environment information.

VIPER Environment File

In order to use VIPER, the information about Kafka Cloud needs to be saved in a file called **VIPER.ENV**. This file contains all the necessary information needed to connect to Kafka Cloud. The details of this file are shown in Table 6-1. Several configuration parameters are available. These configurations are set for each instance of VIPER, VIPERviz, and HPDE. All sensitive information such as passwords will be hidden after the file is saved and accessed by VIPER. If you update this file with new information, you must restart VIPER, VIPERviz, and HPDE for new information to take effect. We will refer back to this table as we build our TML solution. This file is shared between VIPER, HPDE, and VIPERviz.

Table 6-1. *VIPER.ENV Configuration Details*

Configuration Parameter	Description
KAFKA_ADVERTISED_HOST_NAME	Advertised host name in Kafka server properties. THIS IS OPTIONAL.
KAFKA_ZOOKEEPER_CONNECT	Zookeeper host name and port. THIS IS OPTIONAL.
KAFKA_CONNECT_BOOTSTRAP_ SERVERS	Kafka bootstrap servers – separate multiple servers by comma. THIS IS REQUIRED.
KAFKA_ROOT	Kafka root folder. If using Kafka Cloud, simply enter "kafka". THIS IS OPTIONAL.
HPDE_SERVER	HPDE (Hyper-Predictions for Edge Devices) is required for **Auto Machine Learning**. Specify the host where it is installed. THIS IS OPTIONAL.
HPDE_PORT	HPDE listening port. THIS IS OPTIONAL.
SSL_CLIENT_CERT_FILE	SSL certificate file needed if Kafka is SSL/TLS enabled. THIS IS REQUIRED.
SSL_CLIENT_KEY_FILE	SSL certificate key store file needed if Kafka is SSL/TLS enabled. THIS IS REQUIRED.
SSL_SERVER_CERT_FILE	SSL certificate server key file needed if Kafka is SSL/TLS enabled. THIS IS REQUIRED.

(continued)

Table 6-1. (*continued*)

Configuration Parameter	Description
CLOUD_USERNAME	SASL_PLAIN username to connect to Confluent Cloud. THIS IS REQUIRED.
CLOUD_PASSWORD	SASL_PLAIN password to connect to Confluent Cloud. THIS IS REQUIRED.
MAILSERVER	SMTP mailserver host name for sending emails. This is needed if using **AiMS Dashboard** to monitor algorithms in Kafka. THIS IS OPTIONAL.
MAILPORT	SMTP mailserver port for sending emails. This is needed if using **AiMS Dashboard** to monitor algorithms in Kafka. THIS IS OPTIONAL.
FROMADDR	From address to put in the emails. This is needed if using **AiMS Dashboard** to monitor algorithms in Kafka. THIS IS OPTIONAL.
SMTP_USERNAME	SMTP username. This is needed if using **AiMS Dashboard** to monitor algorithms in Kafka. THIS IS OPTIONAL.
SMTP_PASSWORD	SMTP password. This is needed if using **AiMS Dashboard** to monitor algorithms in Kafka and alerts are turned on. THIS IS OPTIONAL.
SMTP_SSLTLS	Mailserver SSL/TLS enabled: true or false. This is needed if using **AiMS Dashboard** to monitor algorithms in Kafka and alerts are turned on. THIS IS OPTIONAL.
SERVICE_USERNAME	If using ServiceNow, specify the ServiceNow web page login username. This is needed if using **AiMS Dashboard** to monitor algorithms in Kafka and alerts are turned on. THIS IS OPTIONAL.
SERVICE_PASSWORD	If using ServiceNow, specify the ServiceNow web page login password. This is needed if using **AiMS Dashboard** to monitor algorithms in Kafka and alerts are turned on. THIS IS OPTIONAL.
SERVICE_ASSIGNEE	If using ServiceNow, specify the ServiceNow name to assign the ServiceNow ticket to. This is needed if using **AiMS Dashboard** and alerts are turned on. THIS IS OPTIONAL.

(*continued*)

Table 6-1. (*continued*)

Configuration Parameter	Description
SERVICE_FORM_FIELDS	`{"key1":"Assignedto","key2":"LastReadofTopic","key3":"Consumerid", "key4":"Brokerhost","key5":"Brokerport","key6":"Companyname", "key7":"Contactemail","key8":"Contactname","key9":"Description", "key10":"Location","key11":"Topic","key12":"Priority","key13":"Producerid","key14":"LastWritetoTopic"}`
	Users should replace the "Key" values with the names of the fields in the ServiceNow Form. VIPER will update the key values when submitting the incident to ServiceNow. This is needed if using **AiMS Dashboard** and alerts are turned on. THIS IS OPTIONAL.
SERVICE_CONTENT TYPE=application/json	ServiceNow web page content type. This can be changed, but **application/json** should be fine. This is needed if using **AiMS Dashboard** and alerts are turned on. THIS IS OPTIONAL.
POLLING_ALERTS	Polling for alerts in minutes. This is needed if using **AiMS Dashboard** and alerts are turned on. VIPER will poll for alerts and wait in minutes for the next poll. THIS IS OPTIONAL.
COMPANYNAME	Specify company name. This is used when sending emails from the AiMS Dashboard. THIS IS OPTIONAL.

For the examples in this book, in your VIPER.ENV file, make the following update (your information will of course be different than shown here; I have put my Kafka cluster address):

- KAFKA_CONNECT_BOOTSTRAP_SERVERS=pkc-419q3.us-east4. gcp.confluent.cloud:9092[7]

---

[7]Note in our example, the parameters required in the document do not need to be string formatted between quotes.

- CLOUD_USERNAME=[Key]

  a.   Replace [Key] with the Key copied in Figure 6-6.

- CLOUD_PASSWORD=[Secret]

  a.   Replace [Secret] with the Secret copied in Figure 6-6.

Because we are encrypting traffic to, and from, Kafka Cloud, we need to set up a two-way encryption using SSL/TLS. To do this, we need to convert three files to PEM format.[8] (Note that example .PEM files are provided in the VIPER, VIPERviz, and HPDE distribution files on GitHub, but these are for testing; they should be replaced with your official file certificates.)

- Client certificate file

- Client key file

- Server certificate file

Note that you may need to have the following files[9] beforehand:

- kafka.server.truststore.jks

- kafka.server.keystore.jks

Here are common steps to do this:

1) First, extract the Certificate Authority (CA) using keytools and openssl:

- **$keytool** -importkeystore -srckeystore kafka.server.truststore.jks -destkeystore server.p12 -deststoretype PKCS12

- **$openssl** pkcs12 -in server.p12 -nokeys -out server.cer.pem

  - **Output:** server.cer.pem

2) Next, convert the client keystore:

- **$keytool** -importkeystore -srckeystore *kafka.server.keystore.jks* -destkeystore client.p12 -deststoretype PKCS12

---

[8]www.ssl.com/guide/pem-der-crt-and-cer-x-509-encodings-and-conversions/#:~:text=PEM%20(originally%20%E2%80%9CPrivacy%20Enhanced%20Mail,%2D%2D%2D%2D%2D%20)

[9]https://docs.oracle.com/cd/E19509-01/820-3503/ggfen/index.html

- **$openssl** pkcs12 -in client.p12 -nokeys -out client.cer.pem

- **$openssl** pkcs12 -in client.p12 -nodes -nocerts -out client.key.pem

  - **Output:** client.cer.pem

  - **Output:** client.key.pem

You will then update the VIPER.ENV file with the full path to the PEM files as follows:[10]

- SSL_CLIENT_CERT_FILE=c:/viper/client.cer.pem

- SSL_CLIENT_KEY_FILE=c:/viper/client.key.pem

- SSL_SERVER_CERT_FILE=c:/viper/server.cer.pem

If everything was done correctly, VIPER, HPDE, and VIPERviz will be able to securely communicate with Kafka using SSL/TLS encryption. The next section discusses how to set up these technologies.

VIPER, VIPERviz, and HPDE Setup

VIPER, VIPERviz, and HPDE can be installed on Windows, Linux, or MacOS operating systems.[11] There are no additional software components or databases needed. Each of these technologies listens for connections on a host IP address and port number. This means you can connect to them using REST API along with the TML Python library[12] (we will use this library in the chapter). Solution architects and developers will also find it easy to use these technologies like *microservices* that can support specific TML solutions, which makes creating very large TML solutions easier for load shedding. All of these technologies are tightly integrated with Apache Kafka, which acts as the data backbone. This allows developers to build TML solutions regardless of the number of data streams, algorithms, and insights consumed. Furthermore, these technologies do

---

[10]This assumes you created a c:/viper folder in Windows. For Linux/Mac, you will need the full path to the PEM files.

[11]You can also view this YouTube video detailing the setup information: www.youtube.com/watch?v=b1fuIeC7d-8

[12]https://pypi.org/project/maadstml/

not require an Internet connection, since Kafka can also be run on-premise (we will use Kafka in the cloud running on Confluent Cloud using Google Cloud Platform). However, if developing production-ready TML solutions, Kafka Cloud is preferred.

VIPER has been verified and tested with Apache Kafka by Confluent, and it can be downloaded from their website[13] or GitHub.[14] Simply unzip the file and follow the installation instructions provided as part of the package. All of the latest versions of the technologies can be downloaded from the GitHub website. When starting VIPER, VIPERviz, or HPDE, users can type "Help" at the command line, and a help screen will detail all the necessary information. Specifically, for users who want to connect via REST API, all of the REST commands will be provided in this screen. For our examples, we will only need to connect to VIPER, and VIPER will manage the connections to HPDE.

If using the MAADSTML Python library, users can write Python code in Jupyter Notebook or by downloading Python.[15] Both Python and Jupyter are very popular, and there are many resources available to you to learn these technologies. Additional questions should be sent to OTICS by emailing info@otics.ca. Or, you can join the Transactional Machine Learning group on LinkedIn.[16] The next section discusses how to create topics and join data streams.

Kafka Topics and Data Streams

A fundamental part of Kafka, and TML solutions, are topics and data streams. We have already discussed topics and data streams in the previous chapters. This section will discuss how we create and use topics (or data streams) to create training datasets. To start, ensure you have Python programming language downloaded and installed. Then, open the Python IDE, or if you use Jupyter Notebook, then start jupyter by invoking *jupyter notebook* in Windows, Linux, or MacOS. We will use Windows OS for our examples. Then, install the MAADSTML Python library by invoking *pip install maadstml in the command prompt*. You are done. Then, to create your topic in Kafka, you will use the function *vipercreatetopic*. This Python function will require the following input parameters:[17]

---

[13]www.confluent.io/hub/oticsinc/maads-viper
[14]https://github.com/smaurice101/transactionalmachinelearning
[15]www.python.org/
[16]www.linkedin.com/groups/13930079/
[17]See https://pypi.org/project/maadstml/ for more details.

```
vipercreatetopic(vipertoken,host,port,topic,companyname,contactname,conta
ctemail,location, description,enabletls=1, brokerhost='',brokerport=-999,
numpartitions=2, replication=3, microserviceid='')
```

where

- vipertoken – This token is included in the distribution VIPER file.

- host – This is the host address for VIPER.

- port – This is the port VIPER is listening on for connections.

- topic – This is the topic you want to create. You can separate multiple topics by a comma.

- companyname – This is the name of your company.

- contactname – Contact name of the person creating the topic.

- contactemail – Email of the person creating the topic.

- location – Location of the person creating the topic.

- description – Description of the topic.

- enabletls – Set this to 1 if SSL/TLS is enabled in Kafka; by default, this is already enabled in Kafka Cloud.

- brokerhost – Specify the Kafka broker host. This is optional; if left empty, then VIPER will use the value in KAFKA_CONNECT_ BOOTSTRAP_SERVERS.

- brokerport – Specify the Kafka broker port. This is optional; if left empty, then VIPER will use the port value in KAFKA_CONNECT_ BOOTSTRAP_SERVERS.

- numpartitions – Specify the number of partitions to create in the topic. While the default is set to 2, you may need to increase this over time if the number of people who want to consume from this topic increases, to ensure parallel processing of information delivery. For example, if you have 10 people who want to consume from this topic, consider setting the partition to 10.

- replication – This is the replication factor that tells Kafka how many times to duplicate the topic for failovers. The server-side minimum for Kafka is 3.

- microserviceid – If using reverse proxy for load shedding, then the name of the proxy can be specified here.

Some may question why the companyname, contactname, and so on are needed. These will become useful for managing topics in AiMS. If you create thousands of topics, having the ability to identify who created the topic and what they are meant for will help to improve the management of the topics (i.e., TML solutions), especially if something goes wrong with the TML solution. You may find it useful to put the business/product name for the companyname to enable corporate IT to better track the TML solution. Once the topic is created, the function will return a ***producerid***. This producerid will be needed to produce data to the topic. Every topic will have a unique producerid.

To produce data to the topic, we will use the function ***viperproducetotopic***. This Python function will require the following input parameters:

```
viperproducetotopic(vipertoken,host,port,topic,producerid,enabletls=1,delay
=100,inputdata='',maadsalgokey='',maadstoken='',getoptimal=0,externalpredic
tion='',brokerhost='',brokerport=-999,microserviceid='')
```

where

- vipertoken – This token is included in the distribution VIPER file.

- host – This is the host address for VIPER.

- port – This is the port VIPER is listening on for connections.

- topic – This is the topic you want to write to. You can separate multiple topics by a comma. If using multiple topics, you must have the same number of producer ids (separated by commas) and same number of externalprediction (separated by commas). Producing to multiple topics at once is convenient for synchronizing the timing of streams for machine learning and to accelerate topic creation for large number of topics.

- producerid – This is the producerid returned from ***vipercreatetopic***.

- enabletls – Set this to 1 if SSL/TLS is enabled in Kafka; by default, this is already enabled in Kafka Cloud.

- delay – This delay parameter is the number of milliseconds that VIPER waits for Kafka to respond with a confirmation that the message was written to the Kafka topic.

- inputdata – For TML, this can be ignored.

- maadsalgokey[18] – For TML, this can be ignored.

- maadstoken – For TML, this can be ignored.

- getoptimal – For TML, this can be ignored.

- externalprediction – Enter a value in this parameter to write to the topic.

- brokerhost – Specify the Kafka broker host. This is optional; if left empty, then VIPER will use the value in KAFKA_CONNECT_ BOOTSTRAP_SERVERS.

- brokerport – Specify the Kafka broker port. This is optional; if left empty, then VIPER will use the port value in KAFKA_CONNECT_ BOOTSTRAP_SERVERS.

- microserviceid – If using reverse proxy for load shedding

You could use the preceding function in a loop to continuously write data to the Kafka topic. You could produce to several topics in parallel, depending only on how you architected your solution and the number of instances of VIPER running. This function will return a JSON object with details on the value written, the key of the value in Kafka, including the partition written to, the offset, and the timestamp. Each data point can be uniquely identified in Kafka by

- Topic

- Partition

- Offset

---

[18]This is used for MAADS deep learning technology, and it is a separate technology that is not covered in this book. For information, users can email info@otics.ca.

Kafka uses unique keys to optimize storage of data. For very large datasets, this will be important for referencing and consuming data quickly from a topic. Names of topics should be chosen carefully as they will identify the dependent and independent variable streams that will be needed when building training datasets. A simple convention is to use [companyid]-[projectid]-[solution name or id]-[input or output stream]. For example, otics-iot-iotfailureprediction-sensorreadings-input and otics-iot-iotfailureprediction-sensorreadings-predictions-output would signify input and output streams for the TML solution iotfailureprediction. This way, it is immediately apparent what topic users need to subscribe to for consumption and visualization. Regardless of the convention you use, it should be consistent.

While Kafka is very clever in its storage of big data using partitions, this also complicates data retrieval. Specifically, Kafka will choose what partition it wants to write to optimize storage. The challenge when retrieving data is you need to do extra processing to determine which data is the *latest* data. For example, if a topic has thirty (30) partitions, Kafka will write your data to any one of these partitions. When consuming these data, you will not know which partition contains the *latest* data. Luckily, you can tell VIPER to search each partition for the latest data using the partition keyword with a value of -1. Because every value stored in Kafka has a date and time value (a human-readable date/time as well as a Unix time), you can then check each value's date and timestamp to determine which one is the *latest* value. While in many cases this is not an issue, in some cases you may want to know the latest value stored in a topic.

To build a training dataset **on the fly**, we will use the following three functions:

- vipercreatejointopicstreams
- viperproducetotopicstream
- vipercreatetrainingdata

First, **vipercreatejointopicstreams** is defined as follows:

```
vipercreatejointopicstreams(vipertoken,host,port,topic,topicstojoin,company
name,contactname,contactemail, description,location,enabletls=0,brokerhost=
'',brokerport=-999, replication=1, numpartitions=1, microserviceid='')
```

where

- vipertoken – This token is included in the distribution VIPER file.
- host – This is the host address for VIPER.

- port – This is the port VIPER is listening on for connections.

- topic – This is the topic that will be created for the joined streams.

- topicstojoin – This is a list of data streams (i.e., dependent and independent variable streams) you want to join. These data streams should exist. Separate multiple streams with a comma.

- Specify companyname, contactname, contactemail, description, location, enabletls, brokerhost, brokerport, replication, numpartitions, and microserviceid, as described earlier, for this topic. Note some fields are optional.

This function will return a ***producerid*** needed to write to this topic. This function acts as a pointer to topics that will be joined. This pointer will be used in the next function that will produce data to this stream. Specifically, once the joined topic stream is created, we can now produce to this stream. To do this, we use the function ***viperproducetotopicstream***:

```
viperproducetotopicstream(vipertoken,host,port,topic,producerid,offset,maxr
ows,enabletls,delay, brokerhost='', brokerport=-999, microserviceid='')
```

- vipertoken – This token is included in the distribution VIPER file.

- host – This is the host address for VIPER.

- port – This is the port VIPER is listening on for connections.

- topic – This is the topic that will contain all the data from the joined streams.

- producerid – This is the producerid returned from ***vipercreatejointopicstreams***.

- offset – This is the location of the data to start from each of the data streams. Specifically, if offset=0, then this function will start from the beginning of each stream it joins. If offset=-1, it will start from the latest value. This parameter is useful to avoid retrieving large amounts of data, which could impact the performance of your TML solution.

- maxrows – This is the number of offsets to ***roll back*** the streams. For supervised learning models, a minimum of 30 data points must be in each stream.

- enabletls, delay, brokerhost, brokerport, and microserviceid are as defined earlier.

This function ***simultaneously*** consumes data from all data streams and writes it to the joined topic stream. Recall that the data streams to join are in the parameter: ***topicstojoin***. Readers can make use of this powerful function to continuously read from multiple data streams and join them in real time. The ***maxrows*** is an offset to roll back the streams. Specifically, as VIPER concurrently retrieves data from each stream, it uses the maxrows offset to roll back each stream, then consolidates all data streams in the stream topic. This function will perform the following steps:

1. Use maxrows to ensure each stream has the same number of values. For example, streams that are joined may have different numbers of data points; for machine learning, we must have the same number of data in each stream.

2. Prune the streams to ensure they all have the same amount of data. For example, if you join three streams (Stream 1 = 100 data points, Stream 2 = 150 data points, Stream 3 = 50 data points), then this function will take the stream with the lowest number (Stream 3) and prune Streams 1 and 2 to 50. This ensures all streams have at most 50 data points.

3. Produce the new, pruned, data streams to a new topic to construct the training dataset.

VIPER will roll back the streams if the offset=-1, meaning that if the offset=-1, you are telling VIPER to go to the ***last offset*** of each stream (technically the end of the stream) and then roll back the stream to an offset determined by rollback=lastoffset-maxrows. This stream, which contains the consolidated data, will be used to construct the training dataset.

Before we create the training dataset, we need to subscribe to the topic just created. To do this, call the function **vipersubscribeconsumer:**[19]

```
vipersubscribeconsumer(vipertoken,host,port,topic,companyname,contactname,
contactemail, location,description, brokerhost='', brokerport=-999,
groupid='',microserviceid='')
```

where

- vipertoken – This token is included in the distribution VIPER file.

- host – This is the host address for VIPER.

- port – This is the port VIPER is listening on for connections.

- topic – Name of the topic to subscribe to

- All other parameters are similar to before.

The function will return a consumer id: ***consumerid***. Every consumer of a topic will use a unique consumer ID. The consumer ID allows AiMS to track, and control, every consumer and record the number of bytes read. We can also create a topic for our training dataset by calling the function ***vipercreatetopic***, which will return the ***producerid***. We will need the producerid and consumerid to create our training dataset.

We can now finally create our training dataset by calling ***vipercreatetrainingdata***:

```
vipercreatetrainingdata(vipertoken,host,port,consumefrom,produceto,de
pendentvariable, independentvariables, consumerid, producerid,company
name,partition=-1, enabletls=0,delay=100, brokerhost='',brokerport=-
999,microserviceid='')
```

- vipertoken – This token is included in the distribution VIPER file.

- host – This is the host address for VIPER.

- port – This is the port VIPER is listening on for connections.

---

[19]There can be many consumers. If you are generating anomaly predictions, your consumers will be interested in seeing which transactions are anomalous or not using VIPER visualization.

- consumefrom – This is the topic we created and produced the joined stream data to using the function ***viperproducetotopicstream***. We also subscribed to this topic by calling ***vipersubscribeconsumer*** and will use the returned consumerid.

- produceto – This is the topic we will produce the training dataset to, and it is the topic we created using the function ***vipercreatetopic***.

- dependentvariable – Name of the dependent variable stream. This should match the dependent variable in *vipercreatejointopicstreams*.

- independentvariables – Name of the independent variable(s) streams. This should match the independent variable(s) in *vipercreatejointopicstreams*.

- consumerid – This is the consumerid for the *consumefrom* topic.

- producerid – This is the producerid for the *produceto* topic.

- offset – This is the offset or location in the *consumefrom* topic to start reading from.

- companyname – Company name of the training dataset owner.

- partition – This is the partition location of the data generated by ***viperproducetotopicstream***. This is important to ensure you are using the right data to create the training dataset.

- enabletls and delay are as described earlier.

- brokerhost, brokerport, and microserviceid are as before.

The training dataset will be stored in Kafka, in the topic specified in ***Produceto***. We are finally ready to call HPDE to perform TML on the training dataset for supervised learning. To do this, we need to call ***viperhpdetraining***:

```
viperhpdetraining(vipertoken,host,port,consumefrom,produceto,companyname,
consumerid,producerid, hpdehost, viperconfigfile, enabletls=1,partition=-
1,deploy=0, modelruns=50, modelsearchtuner=80, hpdeport=-999,offset=-1,
islogistic=0, brokerhost='', brokerport=-999,timeout=120,microserviceid='')
```

where

- vipertoken, host, and port are VIPER connection details as before.

- consumefrom – Is the Kafka topic name we produced the training dataset to in the function *vipercreatetrainingdata* in the parameter *Produceto*. You should have already subscribed to this topic and saved the consumerid.

- produceto – This is the topic name that HPDE will store the details of the optimal algorithm. You should have already created this topic and saved the *producerid*.

- Companyname – Specify your company name.

- Consumerid – The consumerid of the topic in the *consumefrom* parameter.

- Producerid – The producerid of the topic in the *produceto* topic.

- Hpdehost – The host address for HPDE.

- Hpdeport – The port HPDE is listening on.

- Viperconfile – The full path of the VIPER.ENV file on disk.

- Enabletls – Set to 1 for SSL/TLS or 0 if SSL/TLS is not enabled in Kafka.

- Partition – This is the partition number of the training dataset returned by *vipercreatetrainingdata*. This partition ensures you are using the right training dataset.

- Deploy – If 1, then HPDE will save the algorithm in the **deploy** folder. If 0, it will save it to the Models folder only. This is useful if you want to test the algorithm and not use it in production yet.

- Modelruns – The number of models HPDE will run through to find the best algorithm. The higher the number, the longer HPDE will take to find the optimal algorithm.

- Modelsearchtuner – An integer between 0 and 100. This variable will attempt to fine-tune the model search space. A number close to 0 means you will have lots of models, but their quality may be low; a number close to 100 (default=80) means you will have fewer models, but their quality will be higher.

- Offset – The offset number location of the training dataset. This is usually set to -1 to choose the latest training dataset in the last offset.

- Islogistic – If the dependent variable is a binary variable, then set this parameter to 1, otherwise set to 0 for continuous dependent variable.

- Brokerhost, brokerport, and microserviceid are as before.

- Timeout – This is the number of seconds that VIPER will wait for HPDE to respond.

If HPDE responds successfully, then, congratulations, you have just used the core TML functions. We will apply these functions to an actual example to give you a better idea of the use cases that these can be applied to. The training function will return all the information required to make a prediction or do optimization using ***viperhpdepredict*** and ***viperhpdeoptimize*** functions, respectively. An example TML output is shown in Listing 6-1.

Listing 6-1. Example TML Output

```
{
    "Algokey": "ConsumerId-TYq4wDnyS5pVqzImGQIwVQzGD1LrZO_json",
    "Algo": "ConsumerId-TYq4wDnyS5pVqzImGQIwVQzGD1LrZO_jsonrdg",
    "DependentVariable": "viperdependentvariable",
    "Forecastaccuracy": .80,
    "Filename": "./hpdedata/ConsumerId-TYq4wDnyS5pVqzImGQIwVQzGD1LrZO.
csv",
    "Fieldnames": "Date,viperindependentvariable1,viperindependentvariab
le2",
    "TestResultsFile": "./models/ConsumerId-
TYq4wDnyS5pVqzImGQIwVQzGD1LrZO_json_predictions.csv",
    "Deployed": 1,
    "DeployedTo": "Local Machine Deploy Folder",
```

```
    "Created": "2021-01-14T21:15:58.3153622-05:00",
    "ConsumeridFrom": "ConsumerId-TYq4wDnyS5pVqzImGQIwVQzGD1LrZ0",
    "Producerid": "ProducerId-fuj1ygdy-E5noyO2CtKjDNOS6UeaB4",
    "ConsumingFrom": "trainingdata2",
    "ProduceTo": "trainined-params",
    "Companyname": "OTICS",
    "BrokerhostPort": "pkc-419q3.us-east4.gcp.confluent.cloud:9092",
    "Islogistic": 0,
    "HPDEHOST": "192.168.0.13:8001",
    "HPDEMACHINENAME": "Guru",
    "Modelruns": 10,
"Modelsearchtuner": 80,
    "TrainingData_Partition": 0,
    "BytesWritten": 944,
    "kafkakey": "EkFXhJJ3fvTpquCyXEgUq3HZ-CsNnT",
    "Partition": 0,
    "Offset": 4
}
```

The example output is JSON formatted data. It shows keys and values for all important information you need to use for predictions and optimization. This JSON output is described in Table 6-2.

Table 6-2. *JSON Output Keys' Description*

Key	Description
Algokey	A unique key generated by HPDE to identify the algorithm.
Algo	Name of the algorithm that is the best fit to the data.
DependentVariable	Dependent variable stream you chose for the TML model.
Forecastaccuracy	Forecast accuracy of the model in percentage. The forecast accuracy is the MAPE (mean absolute percentage error) and determined by comparing the predictions of the model to the test dataset.
Filename	Physical file of the training dataset. This is useful if you want to audit the (training) data used to determine the best algorithm.

(continued)

Table 6-2. (*continued*)

Key	Description
Fieldnames	Independent variable streams used in the TML model. A Date variable is added as a placeholder but of course ignored in the estimation.
TestResultsFile	This is the physical file of the results from the MAPE calculation.
Deployed	If 1, then this algorithm is deployed to the /deploy folder.
DeployedTo	Location where the algorithm is stored.
Created	Date and time when the model is estimated.
ConsumeridFrom	Consumer ID of the topic you are consuming from in Kafka.
Producerid	Producer ID of the topic you are producing to in Kafka.
ConsumingFrom	Kafka topic containing the training dataset for machine learning by HPDE.
ProduceTo	Kafka topic that the algorithm is produced to.
Companyname	Your company name.
BrokerhostPort	Kafka broker host and port – you can use Kafka on Amazon AWS, Microsoft, Google Cloud Platform, or Confluent Cloud. For this example, we used Google Cloud Platform from Confluent Cloud.
Islogistic	If 1, the model is logistic (with categorical dependent variable); otherwise, the dependent variable is continuous.
HPDEHOST	HPDE host and port.
HPDEMACHINENAME	Machine running HPDE.
Modelruns	The number of models that HPDE will iterate through. The higher this number, the more intensely HPDE will search for the best algorithm or model that best fits the data.
Modelsearchtuner	Value of the model search tuner.
TrainingData_Partition	The partition of the data (generated by ***vipercreatetrainingdata***) that was used for the training dataset. This is useful for auditing training datasets.
BytesWritten	Number of bytes written to the Kafka topic in the ***produceto*** key.
Kafkakey	Kafka key associated with the topic in the ***produceto*** topic.

(*continued*)

Table 6-2. (*continued*)

Key	Description
Partition	Partition number that Kafka has chosen to store the data in the **produceto** topic. This will be important for predictions/optimization to ensure you have the latest location for the algorithm in the Kafka partition.
Offset	Offset number that Kafka has chosen to store the data in the **produceto** topic.

This algorithm is useless if it is not used for predictive or prescriptive analytics. To use it for predictive analytics, we need to call the function ***viperhpdepredict***:

```
viperhpdepredict(vipertoken,host,port,consumefrom,produceto,companyname,co
nsumerid,producerid, hpdehost,inputdata, algokey='',partition=-1,offset=-
1,enabletls=1, delay=1000, hpdeport=-999,brokerhost='', brokerport=-999,
timeout=120, usedeploy=0,microserviceid='')
```

where

- vipertoken, host, and port – Are VIPER connection details.

- consumefrom – Is the topic name in the ***produceto*** parameter in the function ***viperhpdetraining***. You should have already subscribed to this topic and saved the consumerid.

- produceto – Is the topic name you want to write the prediction value to. You should have already created this topic and saved the producerid.

- Companyname – Your company name.

- Consumerid – The consumerid for the topic in the ***consumefrom*** parameter.

- Producerid – The producerid for the topic in the ***produceto*** parameter.

- Hpdehost – Host address for HPDE.

- Hpdeport – Port address for HPDE.

- Inputdata – This is the new raw data for the independent variables. Specifically, these are values for each independent variables that the algorithm will require to make a prediction. The ordering of the values must match the ordering of the independent variables in the optimal algorithm information. Or, you can attach the data stream you produced to in the ***viperproducetotopicstream***; VIPER will continuously read this stream and generate predictions.

- Algokey – This is the key that identifies the optimal algorithm, and it will be returned by ***viperhpdetraining*** in the JSON object.

- Partition – This is the Kafka partition that contains the latest trained algorithm; normally, you can set this to -1 as older algorithms are overwritten.

- Offset – This is the offset location of the optimal algorithm information.

- Enabletls, delay, brokerhost, brokerport, timeout, and microserviceid – Are described earlier.

- Usedeploy – Set to 1, if you want to use the algorithm in the **deploy** (algorithm must exist); otherwise, if 0, it will use the algorithm in the Models folder.

If this function executes successfully, then you have made a prediction from a TML algorithm with new stream data. This function can be called repeatedly to generate predictions. An example output is shown in Listing 6-2.

Listing 6-2. Prediction JSON Output

```
{
    "Hyperprediction": 2.935,
    "Algokey": "ConsumerId-n1NiZUObhCOpIEhAn3q1df3orJpcqy_json",
    "Algo": "ConsumerId-n1NiZUObhCOpIEhAn3q1df3orJpcqy_jsonrdg",
    "Usedeploy": 1,
    "Created": "2021-01-15T16:01:06.9048943-05:00",
    "Unixtime": 1610744466906892800,
    "ConsumeridHPDE": "ConsumerId-c2hTWOANA7DPR-Y6hN6ot2AggzuPfv",
    "Producerid": "ProducerId-HWWwbtbXSykcitWtfN3vY3epHv5OLD",
```

```
    "HPDEHOST": "http://192.168.0.13",
    "HPDEPORT": 8001,
    "Consumefrom": "trainined-params",
    "Produceto": "hyper-predictions",
    "Brokerhost": "pkc-419q3.us-east4.gcp.confluent.cloud",
    "Brokerport": 9092,
    "kafkakey": "4sW8XUFJcGajzBMT3vlxqUONkwqzQC",
    "Inputdata": "146,268",
    "Fieldnames": "Date,viperindependentvariable1,viperindependentvariab
    le2",
    "DependentVariable": "viperdependentvariable",
    "Partition": 2,
    "Offset": 11
}
```

Table 6-3 describes each key in Listing 6-2.

Table 6-3. *TML Prediction JSON Keys' Description*

Key	Description
Hyperprediction	The prediction from the TML algorithm, using the stream input data.
Algokey	Key for the algorithm used to make the prediction.
Algo	The algorithm used to make the prediction. This algorithm is physically on disk and encrypted.
Usedeploy	If 1, HPDE used the algorithm in the /deploy folder; otherwise, it used the algorithm in the Test folder (i.e., /Models folder).
Created	Date and time the algorithm was created.
Unixtime	The Unix time.
ConsumeridHPDE	Consumer ID of the topic HPDE read from to retrieve the algorithm.
Producerid	Producer ID of the topic HPDE wrote the results to.
HPDEHOST	HPDE host IP address.
HPDEPORT	HPDE port.

(continued)

Table 6-3. (*continued*)

Key	Description
Consumefrom	Kafka topic containing the algorithm to use for the hyperprediction.
Produceto	Kafka topic to produce hyperpredictions.
Brokerhost	Kafka broker host.
Brokerport	Kafka broker port.
Kafkakey	Kafka key associated with the prediction.
Inputdata	Values for the independent variables – in the proper order as shown in the **Fieldnames** (excluding Date).
Fieldnames	Name of fields in the TML model.
DependentVariable	Dependent variable for the TML model.
Partition	Kafka partition that contains the Hyperpredictions.
Offset	Kafka Offset – specifying the offset of the hyperpredictions.

We can even go a step further and perform optimization on the optimal algorithm to find the optimal values of the independent variables, by maximizing or minimizing an objective function. For instance, to do this, we call ***viperhpdeoptimize***:

```
viperhpdeoptimize(vipertoken,host,port,consumefrom,produceto,companyna
me,consumerid,producerid, hpdehost,partition=-1,offset=-1,enabletls=0,
delay=100,hpdeport=-999,usedeploy=0, ismin=1,constraints='best',
stretchbounds=20, constrainttype=1,epsilon=10,brokerhost='',brokerpo
rt=-999, timeout=120,microserviceid='')
```

where

- Vipertoken, host, and port – Are VIPER connection details.

- consumefrom – Is the topic name in the ***produceto*** parameter in the function ***viperhpdetraining***.

- Produceto – Is the topic name that you want to save the optimal values to.

- Companyname – Is the name of your company.

- Consumerid – Is the consumerid of the topic in **consumefrom**.

- Producerid – Is the producerid of the topic in **Produceto**.

- Hpdehost, hpdeport – Are the HPDE connection details.

- Partition, offset – Are the location of the optimal algorithm information.

- Enabletls, delay, brokerhost, brokerport, timeout, microserviceid – Are as described before.

- Ismin – Set this to 1 if you want to minimize the dependent variable, 0 if you want to maximize it.

- Constraints – You can leave this at "best" to let VIPER decide; alternatively, these can be set for each of the independent variables by using the format: varname1:min:max,varname2:min:max,…

- Stretchbounds – This is a percentage to stretch the bounds on the independent variable constraints. For example, if stretchbounds=20, and a and b are bounds on variable **var**, then a < **var** < b, then HPDE will decrease a = a-a*0.2 and increase b=b+b*0.2 by 20%.

- Constrainttype – If 1, then HPDE uses the min/max of each variable for the bounds; if 2, HPDE will adjust the min/max by their standard deviation; if 3, then HPDE uses stretchbounds to adjust the min/max for each variable.

- Epsilon – This is a percentage. If epsilon is 10, then once HPDE finds a good local minima/maxima, it uses this epsilon value to perturb the input values by 10% for the independent variables to find the global minima/maxima to ensure you have the best values of the independent variables that minimize or maximize the dependent variable.

If this function executes successfully, then you have performed optimization on this algorithm and will produce optimal values for the independent variables that minimize or maximize the dependent variable. The output is shown in Listing 6-3.

Listing 6-3. TML Optimization Output

```
{
    "UserDetails": {
        "CreatedOn": "Fri, 15 Jan 2021 16:01:16 EST",
        "Unixtime": 1610744476871555800,
        "HPDEHOST": "192.168.0.13:8001",
        "ConstraintType": "USE min/max for bounds",
        "Epsilon": 20,
        "StretchBounds": 10,
        "Usedeploy": 0
    },
    "OptimizedValues": {
        "Objective": "Maximization",
        "ObjectiveFunctionValue": 9.999,
        "OptimalValues": [{
            "viperindependentvariable1": 630.350
        }, {
            "viperindependentvariable2": 780.078
        }]
    },
    "Constraints": [{
        "Max": 854.744,
        "Min": 569.83,
        "Variable": "viperindependentvariable1"
    }, {
        "Max": 837.678,
        "Min": 558.452,
        "Variable": "viperindependentvariable2"
    }],
    "DescriptiveStats": [{
        "Max": 984,
        "Mean": 505.203,
        "Min": 0,
        "STD": 269.828,
        "Variable": "viperindependentvariable1"
```

```
    }, {
        "Max": 994,
        "Mean": 488.21,
        "Min": 1,
        "STD": 290.087,
        "Variable": "viperindependentvariable2"
    }],
    "AdditionDetails": {
        "Consumerid": "ConsumerId-c2hTWOANA7DPR-Y6hN6ot2AggzuPfv",
        "Producerid": "ProducerId-xTo2TFMGy3-HPaVEkBFE581tbwkxU1",
        "Consumefrom": "trainined-params",
        "Produceto": "hpde-optimal-parameters",
        "Brokerhost": "pkc-419q3.us-east4.gcp.confluent.cloud",
        "Brokerport": 9092,
        "BytesWritten": 1007,
        "kafkakey": "uQH9JI99rr28igMUdvFCV4KrPNgrMP",
        "Partition": 0,
        "Offset": 4
    }
}
```

Table 6-4 describes the keys in Listing 6-3.

Table 6-4. *Optimization Output Keys' Description*

Keys	Description
UserDetails	User optimization details.
CreatedOn	Date and time optimization was generated.
Unixtime	Unix time when optimization results generated.
HPDEHOST	HPDE host.

(continued)

Table 6-4. (*continued*)

Keys	Description
ConstraintType	The type of constraints used. There are three types of constraints. If ConstraintType=1, then the independent variables will be bound by their minimum and maximum values. For example, for independentvariable1, the bound is min <=independentvariable1<= max. If ConstraintType=2, then the independent variables will be bound by their minimum value minus their standard deviation. For example, for independentvariable1, the bound is min-std <= independentvariable1<= max-std. If ConstraintType=3, then the independent variables will be bound by their minimum value minus the value in stretchbounds. For example, for independentvariable1, the bound is min-min*stretchbounds<= independentvariable1<= max-max*stretchbounds.
Epsilon	Once HPDE finds a good local minima/maxima, it then uses this epsilon value to find the global minima/maxima to ensure you have the best values of the independent variables that minimize or maximize the dependent variable.
StretchBounds	A percentage number ranging from 0 to 100. This value will be used in the constraints to stretch the bounds, as shown before.
Usedeploy	If 1, use the deployed algorithm (in the /deploy folder); otherwise, use the test algorithm (in the /models folder).
OptimizedValues	These are the optimized values for the objective function and independent variables from the optimization.
Objective	This will indicate your choice between maximization and minimization.
ObjectiveFunctionValue	This is the optimized value of the objective function.
OptimalValues	These are the optimized values for the independent variables from the optimization.
Constraints	This will indicate the constraint bounds for the independent variables.

(*continued*)

Table 6-4. (*continued*)

Keys	Description
DescriptiveStats	This will show descriptive statistics for the independent variables: Min, Max, Mean, STD.
AdditionDetails	These are additional details from the optimization.
Consumer ID	Consumer ID of the topic you are consuming from in Kafka.
Producer ID	Producer ID of the topic you are producing to in Kafka.
Consumefrom	Topic you are consuming from in Kafka.
Produceto	Topic you are producing to in Kafka.
Brokerhost	Kafka broker host.
Brokerport	Kafka broker port.
BytesWritten	The number of bytes written to in Kafka.
Kafkakey	The Kafka key associated with the data stored in Kafka **produceto** topic.
Partition	The partition Kafka wrote the data to in **produceto** topic.
Offset	The offset Kafka generated for the data in **produceto** topic.

The combination of ***viperhpdetraining*** and ***viperhpdeoptimize*** can be powerful to generate outcomes from TML models that can be used for better and faster decision-making. The preceding functions will perform TML on any data streams. Using the optimal algorithms to predict and optimize the values of the independent variables, users can provide advanced machine learning insights in a frictionless manner. A typical TML process flow of, generally, when to use these functions is shown in Figure 6-9 for easier referencing.

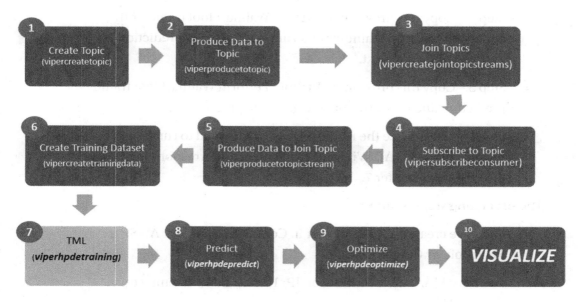

Figure 6-9. *TML Flow Diagram for Supervised Learning*

We will show how this process is applied to the example code provided as part of this book. The next sections will discuss the example code and how they fit into the preceding process.

TML Example Code

As part of this book, you are provided with example TML code. The code is found on GitHub.[20] The code is divided in two sets:

CODE SET 1: This set of programs will go through an example of predicting and optimizing foot traffic at ~11,000 Walmart stores. To run this example, follow the following steps:

- Step 1 – Copy this program in Python: Produce Walmart data to the Kafka cluster and let it run for 5 minutes or so and then run the machine learning code next.

---

[20]https://github.com/smaurice101/transactionalmachinelearning

- Step 2 – Copy this program in Python: Walmart foot traffic TML and let this run for 5 minutes or so and then run the prediction/optimization code next.

- Step 3 – Copy this program in Python: Perform Walmart foot traffic prediction and optimization and generate predictions.

- Step 4 – To visualize the results in step 3, you need to run MAADS Visualization (MAADSViz) and then enter the corresponding URLs discussed in Chapter 7.

The preceding steps assume

1. You have created a Kafka cluster in Confluent Cloud (or AWS, Microsoft, or Google Cloud).

2. You have MAADSViz running on IP: 127.0.0.1 and listening on Port: 8003.

3. You downloaded views zip and extracted contents to the viperviz/views folder.

CODE SET 2: This set of programs will perform bank fraud detection on 50 bank accounts and 5 fields in each transaction. It will detect fraud in real time by following the steps:

- Step 1 – Copy this program in Python: Produce bank account data to the Kafka cluster and let it run for 5 minutes or so and then run the anomaly detection code next.

- Step 2 – Copy this program in Python: Perform transactional bank fraud detection on streaming data and let it run to produce anomaly results.

- Step 3 – Visualize anomalies (discussed in Chapter 7).

The preceding steps assume

1. You have created a Kafka cluster in Confluent Cloud (or AWS, Microsoft, or Google Cloud).

2. You have MAADSViz running on IP: 127.0.0.1 and listening on Port: 8003.

3. You downloaded views zip and extracted contents to the viperviz/views folder.

It should be noted that you should monitor your cloud billing/payments during and after the running of the preceding programs. Specifically

- DELETE YOUR CLUSTER WHEN YOU ARE DONE.

- DO NOT LET YOUR CLUSTER RUN IF YOU ARE NOT USING IT.

- The preceding programs will auto-create all data very quickly. So you can DELETE your cluster immediately.

- Confluent will give you $200 free cloud credits.[21] The preceding programs will consume a fraction of this free amount.

The next section discusses the code set 1: Walmart foot traffic prediction.

Walmart Foot Traffic Prediction and Optimization with TML

The best way to get familiar with TML is to see it in action. The simple Walmart example using Figure 6-9 will do just that. Each step in the process is a core step, but you can easily change the Python code to fit your needs. It will be useful for you to see how the three programs connect with the visualization to solve a particular use case of predicting foot traffic in Walmart locations around the world.[22] There are three program files in the Walmart example:

1. Produce simulated data streams to the Kafka cluster for the[23]

    - Dependent variable – Foot traffic

    - Independent variables – Hour of day, month of year, Walmart store location number (there are 11,000 locations)

---

[21]These credits may vary over time.
[22]This example is found on GitHub.
[23]This file is called: produce data to Kafka cluster.

2. Perform TML on the four data streams: foot traffic, hour of day, month of year, Walmart store locations.[24]

3. Generate insights:[25]

- Predict foot traffic in all 11,000 store locations based on the hour of day, month of year, and store location number.

- Optimize (maximize) foot traffic by finding the optimal values for the hour of day, month of year, and store location. This will allow you to plan how many people you hire at a store location, the amount of inventory, when the best time is to promote products at particular locations, and so on.

4. Visualize the insights for predictions and optimization. Visualization will be discussed in detail in Chapter 7. For now, you should know that you can visualize streaming insights by simply using VIPER visualization and pointing your browser to the HTML files provided for

- Prediction – Use file **prediction.html**.

- Optimization – Use file **optimization.html**.

The process steps in Figure 6-9 are explained as follows with the TML Python functions:[26]

1. Create topics in the Kafka cluster to store your data streams. You create topics using the vipercreatetopic Python function. Topics will be created with the number of partitions of your choosing. In the examples, topics are created with one partition. You should note that partitions are priced higher than data, so you need to allocate partitions carefully. Topics can also be created concurrently, which is useful if you have thousands of topics and want them created very quickly.

---

[24]This file is called: Walmart Foot Traffic TML.

[25]This file is called: Walmart-Predict-and-Optimize-Foot-Traffic.

[26]All details on the TML Python library functions can be found here: https://pypi.org/project/maadstml/

- Code example:[27]

  - streams=["otics-tmlbook-walmartretail-foottraffic-prediction-foottrafficamount-input","otics-tmlbook-walmartretail-foottraffic-prediction-hourofday-input","otics-tmlbook-walmartretail-foottraffic-prediction-monthofyear-input","otics-tmlbook-walmartretail-foottraffic-prediction-walmartlocationnumber-input"]

  - topicnames=','.join(streams)

  - description="TML Book example prediction and optimization modeling"

  - result=maadstml.vipercreatetopic(VIPERTOKEN,VIPERHOST, VIPERPORT,topicnames,companyname,myname,myemail, mylocation,description,enabletls,brokerhost,broker port,numpartitions,replication,microserviceid)

  - topicnames is a list of the topics we will create in the streams array.

2. Produce to Topics with your data. In the example, we simulate the data. Data are produced to each topic very quickly using the function viperproducetotopic. You should dedicate an instance of VIPER on a different port just to handle data production for maximized performance. Data can also be produced concurrently to all topics at the same time. The example produces data to the four data streams concurrently.

   - Code example:

     - topicbuf=','.join(topiclist)

     - produceridbuf=','.join(producerids)

     - delay=7000

---

[27]https://github.com/smaurice101/produce_data_to_kafka

- result=maadstml.viperproducetotopic(VIPERTOKEN, VIPERHOST,VIPERPORT,topicbuf,produceridbuf,1,delay,'',' ', '',0,inputbuf), where

- topicbuf – Is a list of topics

- produceridbuf – Is a list of producer ids for each topic

- inputbuf – Is the value you want to store in each topic

- delay – Of 7000 milliseconds in case Kafka does not respond, then VIPER will wait a maximum of 7 seconds before backing out

3. Join the Topics creates a template topic for the joined topics: foot traffic, hour of day, month of year, and store location number. This template is created using the function vipercreatejointopicstreams. You can easily join any number of topics with this function. This makes TML solutions elastic, because you can create very large or very small solutions with one function.

   - Code example:[28] streamstojoin=["otics-tmlbook-walmartretail-foottraffic-prediction-foottrafficamount-input","otics-tmlbook-walmartretail-foottraffic-prediction-hourofday-input","otics-tmlbook-walmartretail-foottraffic-prediction-monthofyear-input","otics-tmlbook-walmartretail-foottraffic-prediction-walmartlocationnumber-input"]

   - streamstojoin=','.join(streamstojoin)

   - result=maadstml.vipercreatejointopicstreams (VIPERTOKEN,VIPERHOST,VIPERPORT,joinedtopic, streamstojoin, companyname, myname, myemail, description, mylocation, enabletls, brokerhost, brokerport, replication, numpartitions, microserviceid)

---

[28]https://github.com/smaurice101/Walmart-Foot-Traffic-Transactional-Machine-Learning

- joinedtopic – otics-tmlbook-walmartretail-foottraffic-prediction-joinedtopics-input – This is the template topic

- streamstojoin – Group of topics to join

4. Subscribe to Topics to consume data from the joined topic stream. Subscription is handled automatically using the function vipersubscribeconsumer. Subscribing to topics is only done once and helps to keep track of all topics. In the example, you subscribe to a topic to consume its data.

- Code example: result=maadstml.vipersubscribeconsumer(VIPE RTOKEN,VIPERHOST,VIPERPORT, joinedtopic, companyname, myname, myemail, mylocation, description, brokerhost,brokerpo rt,groupid,microserviceid)

  - joinedtopic – This is the topic to subscribe to.

5. Produce Data to Join Topics once the joined template is created. You use the function viperproducetotopicstream. This is a powerful function. This is where you roll back the data streams to create a historical dataset for transactional machine learning. Specifically, this function will read the topics you want to join, then connect to each data stream, roll back each stream, and produce the consolidated data concurrently to the joined topics. This process is very fast, and it is a critical part of TML solutions.

- Code example: result=maadstml.viperproducetotopicstream(V IPERTOKEN,VIPERHOST,VIPERPORT, joinedtopic, producerid, startingoffset, rollbackoffsets, enabletls, delay, brokerhost, brokerport,microserviceid)

  - joinedtopic – Produce the stream to this topic

  - rollbackoffsets – The number of offsets to roll back the data stream

  - startingoffset – Set to -1 to go to the end of streams and roll back

6. Create the Training Dataset for TML. You create training datasets using the function vipercreatetrainingdata. This function will read the consolidated data and transform it by ensuring all streams have the same number of rows, then convert the JSON to an array for machine learning. It will also separate the dependent variable from the independent variables. In the example, we use the foot traffic stream as the dependent variable and the hour of day, month of year, and store location number as the independent variable streams.

- Code example:

- result=maadstml.vipercreatetrainingdata(VIPERTOKEN, VIPERHOST,VIPERPORT, consumefrom, producetotopic, dependentvariable, independentvariables, consumerid, producerid, companyname, partition,

  - enabletls, delay ,brokerhost, brokerport, microserviceid)

  - consumefrom – The joinedtopic stream

  - producetotopic – The training dataset to a topic

  - dependentvariable – "otics-tmlbook-walmartretail-foottraffic-prediction-foottrafficamount-input"

  - independentvariables – ["otics-tmlbook-walmartretail-foottraffic-prediction-hourofday-input", "otics-tmlbook-walmartretail-foottraffic-prediction-monthofyear-input", "otics-tmlbook-walmartretail-foottraffic-prediction-walmartlocationnumber-input"]

7. TML to find the best algorithm. By using the function viperhpdetraining, you connect to HPDE which will connect to the training dataset in Kafka and perform TML. It will iterate through seven sets of algorithms:

- Neural networks

- Ridge regression

- Gradient boosting

- Genetic algorithm

- Linear gradient

- Multiple linear regression

- Logistic regression (if the dependent variable is a classification)

Each of these algorithms will also perform hyper parameter tuning on each model. The number of models is specified by the modelruns field. In the example, we have set this to 20. You can increase or decrease this number, but the higher the number, the longer it will take to find the optimal algorithm. Another parameter of interest in the function is the field modelsearchtuner: this field ranges from 0 to 100. A value close to zero will give you lots of models, but their quality may be low; a value close to 100 will give you fewer models, but their quality will be higher. We have set this value to 85 in the example. To see the final set of algorithms, you can go to the /models folder, and it will show you a CSV file, info file, a param file, and the chosen algorithm. The most interesting is the CSV file which shows the result of using the algorithm on the holdout dataset. It shows the Actual Y (dependent variable) vs. Predicted Y (dependent variable) and allows you to see how well the algorithm performs in predicting the holdout dataset. The param file will record the last set of parameters in the algorithm that resulted in the best model. For your next retraining, HPDE uses these parameters and tries to find better models, if any. If a better model is not found, it stays with the older parameters. HPDE uses MAPE (mean absolute percentage error) to determine which model or algorithm is better than a competing model. The info file contains the details on the best model, and every time a new (better) model is found, this file is updated. With HPDE, you can easily retrain models on new streaming data automatically, which leads to a frictionless machine learning process.

- Code example:[29]

- consumefrom="otics-tmlbook-walmartretail-foottraffic-prediction-trainingdata-input"

- producetotopic="otics-tmlbook-walmartretail-foottraffic-prediction-trained-params-input"

- deploy=1

- modelruns=20

- offset=-1

- islogistic=0

- networktimeout=600

- modelsearchtuner=85

- result=maadstml.viperhpdetraining(VIPERTOKEN,VIPERHOST,VIPERPORT, consumefrom, producetotopic, companyname, consumeridtrainingdata2, producerid, hpdehost, viperconfigfile, enabletls, partition_training, deploy, modelruns, modelsearchtuner, hpdeport,offset,islogistic, brokerhost, brokerport, networktimeout, microserviceid)

  - consumefrom – Consume from the training dataset topic.

  - producetotopic – Produce the trained model to this topic.

  - deploy – If deploy=1, the model is deployed to the /deploy folder; otherwise, it is in the /models folder. This is useful for testing the model before using it in production.

  - modelruns – Number of model iterations. You can increase this number to find better models, but it will take longer.

  - offset – offset=-1 means to go to the end of the consume from stream to get the data for training.

  - islogistic – If islogistic=0, then the model is not a logistic model; if islogistic=1, then it is.

---

[29]https://github.com/smaurice101/Walmart-Foot-Traffic-Transactional-Machine-Learning

- networktimeout – This is the maximum amount in seconds that VIPER will wait for HPDE to finish before backing out.

- modelsearchtuner – This number will fine-tune the model search space. If the number is close to 0, you will get more models to choose from, but the quality may be low; if the number is close to 100, then you will get fewer models to choose from, but their quality will be higher. Note that only one optimal model is chosen out of all competing models based on the highest MAPE value.

- partition_training – This is the partition that contains the data for training. This is an important parameter and ensures you are using the right dataset.

8. Predict values with the trained model. HPDE uses the details in the info file to determine which algorithm to use to predict values of the dependent variable. By using the function viperhpdepredict, you can predict values of the dependent variable very quickly. For our foot traffic example, we will predict how much foot traffic occurs at Walmart locations in a given hour, month, and store location.

- Code example:[30]

    - producetotopic="otics-tmlbook-walmartretail-foottraffic-prediction-results-output"

    - inputdata=joinedtopic

    - consumefrom="otics-tmlbook-walmartretail-foottraffic-prediction-trained-params-input"

    - mainalgokey=""

    - offset=-1

    - delay=60000

    - usedeploy=1

---

[30]https://github.com/smaurice101/Walmart-Predict-and-Optimize-Foot-Traffic

- networktimeout=120

- maadstml.viperhpdepredict (VIPERTOKEN,VIPERHOST, VIPERPORT ,consumefrom, producetotopic, companyname, consumeridtraininedparams, produceridhyperprediction, hpdehost, inputdata, mainalgokey, -1, offset, enabletls, delay, hpdeport, brokerhost, brokerport, networktimeout, usedeploy ,microserviceid)

  - ***producetotopic*** – The predictions will be produced to this topic. *This is the topic you would use for visualization.* You can subscribe to this topic and generate a consumer id that can be used in the visualization.

  - ***inputdata*** – This points to an input data stream. This is the same topic that is used in the *viperproducetotopicstream* and contains the new input data that will be used by the model to predict the dependent variable.

  - ***consumefrom*** – This is the topic that contains the trained parameters (algorithm).

  - ***mainalgokey*** – If this is empty, then the algorithm stored in the consumefrom topic is used. This is simply for performance: if you specify the name of the algorithm here, then the prediction function does not have to look into the Kafka topic for the algorithm. You can find the name of the algorithm in the ***"Algokey"*** field in the file suffixed by ***"_.info"*** in the ***/models*** folder; the prefix of this file is the name of consumerid.

  - ***offset*** – The offset of the trained parameters. Note: The ***"_.info"*** file is the metadata for the details of the trained model. As you retrain the model, the "Algo" field in this meta file may point to a new (better) algorithm, but the "_.info" file will ***not*** change.

- *delay* – This is the maximum delay in milliseconds that VIPER will wait for Kafka to return back a successful response before backing out.

- *usedeploy* – This tells HPDE to use the model in the /deploy folder when usedeploy=1 or in the /models folder when usedeploy=0.

- *networktimeout* – This is the number of seconds that VIPER will wait for HPDE to finish its tasks before backing out.

9. Optimize the dependent variable by finding the optimal values of the independent variables. By using the function viperhpdeoptimize, you can quickly find optimal values. In the foot traffic example, you will find the best values of the independent variables that maximize Walmart foot traffic.

- Code example:[31]

  - *consumefrom*="otics-tmlbook-walmartretail-foottraffic-prediction-trained-params-input"

  - *delay*=10000

  - *offset*=-1

  - *ismin*=0

  - *constraints*='best'

  - *stretchbounds*=20

  - *constrainttype*=1

  - *epsilon*=20

  - *timeout*=120

- maadstml.viperhpdeoptimize(VIPERTOKEN,VIPERHOST,VIPERPORT,consumefrom,producetotopic, companyname,consumeridtraininedparams,

---

[31]https://github.com/smaurice101/Walmart-Predict-and-Optimize-Foot-Traffic

- producerid,hpdehost,-1, offset, enabletls, delay, hpdeport, usedeploy, ismin, constraints, stretchbounds, constrainttype, epsilon, brokerhost, brokerport, timeout, microserviceid)

- *consumefrom* – Topic to consume from. This topic contains the trained parameters. We are going to predict values for the dependent variable that we want to optimize.

- *delay* – Delay in milliseconds that VIPER will use to wait for Kafka to respond before backing out.

- *offset* – If offset=-1, then HPDE looks to the end of the consumefrom stream for the algorithm.

- *ismin* – If ismin=1, then you want to minimize the objective function (which is the dependent variable equation that was determined when you trained your model).

- *constraints* – You can specify "best" to let HPDE determine the constraints, or you can specify custom constraints. Custom constraints must be in the following format: varname1:min:max,v arname2:min:max,...

- *stretchbounds* – This is the value between 0 and 100 and specifies the percentage to stretch the lower and upper bounds of your constraints.

- *constrainttype* – If 1, then HPDE uses the min/max of each variable for the bounds; if 2, HPDE will adjust the min/max by their standard deviation; if 3, then HPDE uses stretchbounds to adjust the min/max for each variable.

- *epsilon* – Once HPDE finds a good local minima/maxima, it then uses this epsilon value, which is between 0 and 100, to find the global minima/maxima to ensure you have the best values of the independent variables that minimize or maximize the dependent variable.

10. Visualize the predictions and optimization results. Visualization is an important aspect of TML solutions. You can visualize the results of the predictions and optimization immediately as the

insights are written to Kafka. VIPER visualization uses WebSocket through the HTML files provided to push insights to your browser. We discuss visualization in detail in Chapter 7.

The next example program (code set 2) shows how you can perform anomaly detection on bank transactions. Before we apply anomaly detection, let's look deeper into how TML performs unsupervised learning for anomaly detection.

Unsupervised Learning for Anomaly Detection

Unsupervised learning is useful when it is impossible to create a training dataset. For example, in real-time fraud detection, when you cannot easily label the dependent variable, you *do not* have, or know how to classify, a dependent variable. In contrast with supervised learning, machines learn the relationships between the dependent variable and independent variable(s). The machine then captures these learnings in the estimated parameters, that is, a, b, and c, in our weather temperature example. But, for use cases that require anomaly, or outlier, detection in data streams, using supervised learning is impractical. This is because supervised learning requires a dependent variable that has *classifications* of *past* anomalies or outliers in the data. Knowing past anomalies in data streams, then building a dependent variable with this classification, is impossible to do in real time [Bolton & Hand, 1999]. For this reason, using unsupervised learning that does *not* require knowledge of past anomalies and does *not* require a dependent variable is the right approach [Bolton & Hand, 1999].

With TML unsupervised learning, you can now perform anomaly detection and predict whether a transaction is normal or abnormal: predict the probability of likelihood of anomalies. There are a large number of use cases that can make use of this unsupervised learning technique to predict probabilities of anomalies in data streams such as

1. Fraud prediction in financial transactions

2. Product or equipment failure prediction for IoT devices

3. Product recommendation prediction for online products

4. And so on

In the first two cases, the rapid identification of abnormal data reduces the risk of a large financial loss. In the third example, TML exposes latent opportunities to maximize revenue.

As discussed earlier, building TML models for anomaly prediction involves a similar process to that of supervised learning, with few differences:

1. You do not create a training dataset; rather, you will now create a peer group that captures "normal" behaviors in sliding windows from data streams.

2. You will use the peer group as a comparison group to detect "abnormal" behaviors in new data (transactions).

3. You will generate a risk score ranging between 0 and 1, inclusive, that will indicate if the new transaction(s) is "normal" or "abnormal."

The ability for TML to perform anomaly detection using unsupervised learning in real time with data streams is a very powerful technique that can scale to millions or billions of transactions quickly.

TML models for anomaly detection use two MAADSTML Python functions:

1. ***viperanomalytrain***

   a. Creates a dataset from the joined transactions' streams and **rolls back** the streams.

   b. From the dataset, finds the **peer group of transactions** (training dataset) by detecting "normal" transaction behaviors.

   c. Stores the peer group (training dataset) in another Kafka topic.

   d. You can choose parameters to adjust the sensitivity of the algorithm that chooses "normal" behaviors.

2. ***Viperanomalypredict***

   a. Tests new transactions against the peer group of "normal" behaviors.

   b. Does a field-level test for potential anomalies using **advanced algorithms in HPDE**.

c. Predicts the likely anomaly for every transaction in real time.

d. You can choose parameters to adjust the sensitivity of the algorithm that chooses anomalous behaviors.

Given the comprehensive nature of these functions, they will be discussed in detail. First, Table 6-5 describes **viperanomalytrain**:

```
viperanomalytrain(vipertoken,host,port,consumefrom,produceto,produce
peergroupto,produceridpeergroup,consumeridproduceto, streamstoanalys
e,companyname,consumerid,producerid,flags,hpdehost,viperconfigfile,
enabletls=1,partition=-1,hpdeport=-999,brokerhost='',brokerport=9092,
delay=1000,timeout=120,microserviceid='')
```

Table 6-5. *Viperanomalytrain Description*

Variable	Description
vipertoken	VIPER token that is required to start VIPER.
host	Host address where VIPER is listening for connections.
port	Port address where VIPER is listening for connections.
consumefrom	Topic to consume from. This is the topic that ***viperproducetotopicstream*** produced to.
produceto	This topic will contain the formatted data for anomaly training. Consider this as an intermediary data storage.
producepeergroupto	Topic to produce the peer group to.
produceridpeergroup	Producer ID for the peer group topic. Get this from the ***vipercreatetopic*** function.
consumeridproduceto	Consumer ID for the *produceto* topic. Get this from the ***vipersubscribeconsumer*** function.
streamstoanalyse	Names of data streams. Separate multiple streams with a comma. This indicates what streams you are analyzing for anomalies.
companyname	Your company name.
consumerid	Consumer ID for the **consumefrom** topic.

(continued)

Table 6-5. (*continued*)

Variable	Description
producerid	Producer ID for the **produceto** topic.
flags	This is an important variable to control how peer groups are chosen. Specifically, to choose a good peer group of "normal" behaviors, you need to remove any "abnormal" values, because these are the values you want to detect. But every data stream is different, and so every peer group will be different. You do not want to be too restrictive or too relaxed in defining "normal" behavior, because if you accuse a transaction of fraud, when in fact it is not, then this could cause issues with your customers. Therefore, the intention is to minimize the occurrence of false positives and false negatives in your outcomes by choosing the most reasonable representation of normal behaviors among the peers. The flags variable allows you to experiment with different values until you are satisfied that the peer group you have is a reasonably well representation of normal behavior for your data. You must set flags for each data stream specified in **streamstoanalyse**. Separate multiple streams with the symbol ~. Flags must have the following format for **numeric streams**: [**topic name**],[**topictype**= numeric], [**threshnumber**=this number is used to determine whether a numeric value or string value is normal or not. Usually, a number below 0.2200 is normal, but you can experiment with this value],[**lag**=this is used to smooth the function before performing tests for normality; a value of 5 is usually fine], [**zthresh**=standard deviation of the data from the centroid of the data stream; usually a value of 2.5 standard deviations is fine],[**influence**=normal influence on the data is close 1, but depends on the data; usually a value of 0.5 is fine].

(continued)

Table 6-5. (*continued*)

Variable	Description
	Flags must have the following format for **text streams**: [topic name], [topictype=string], [**threshnumber**=this is a threshold to detect similarity in text data. For example, the text values "dog" "dogged" will be tested for similarity using text analytics. A threshold value of 0.810 may be fine but you need to experiment with these values].
	Let's take an example for five data streams that are a combination of numeric and text streams:
	1. viperdependentvariable
	2. viperindependentvariable1
	3. viperindependentvariable2
	4. textdata1
	5. textdata2
	The flags can be set as the following based on the data streams; these values will likely change for different data streams:
	topic=viperdependentvariable, **topictype**=numeric, **threshnumber**=300.15, **lag**=5, **zthresh**=2.5, **influence**=0 .5~**topic**=viperindependentvariable1, **topictype**=numeric, **threshnumber**=0.15300, lag=5, **zthresh**=2.5,
	influence=0.5~topic=viperindependentvariable2, **topictype**=numeric ,**threshnumber**=0.18300,**lag**=5,**zthresh**=2.5,
	influence=0.9~**topic**=textdata1,**topictype**=string,**threshnumber**=10 0.85~**topic**=textdata2,**topictype**=string, **threshnumber**=0.80
hpdehost	Host IP address of HPDE.
Viperconfigfile	Full path address for VIPER configuration file.

(*continued*)

Table 6-5. (*continued*)

Variable	Description
EnabletIs	If 1, then Kafka is SSL/TLS enabled, and VIPER/HPDE will automatically switch to SSL/TLS encryption, otherwise no SSL/TLS.
Partition	This is the partition that the function **viperproducetotopicstream** stored the joined stream to. Specifically, because Kafka stores data in the partition of its choosing, with this variable you can tell Kafka to read the last entry of data in this partition. This ensures that you are using the right data for peer groups.
Hpdeport	Port number of HPDE.
Brokerhost, brokerport, delay, timeout, microserviceid	Same description as before.

All of this is achievable in a single line of code. You may sense that anomaly detection using TML is complicated, but it is actually quite simple. You need to only write one line of code to detect anomalies in any number of combined data streams. For example, in our bank fraud example, say you make a purchase at Walmart, at time of purchase, the point of sale system (POS) will likely record

1. Your name (YN)

2. Date/time of your purchase (DT)

3. Your location (L)

4. Product you purchased (PP)

5. Price you paid (Pr)

6. Method of payment (MP)

7. Store location (S)

There is probably more information that is recorded, but for the sake of example, we will use these seven field types and call them F (=7). Also, I have put variables by each item represented by N, DT, L, PP, Pr, S, and M. Each transaction will be comprised of these record types: $T_i = \{YN_i, DT_i, L_i, PP_i, Pr_i, MP_i, S_i\}$ is transaction $i=1,...N$, where N is a large number. Now, if I am working at a bank and asked you to check every transaction for every account holder, **j=1...M**, for fraud, then I need to analyze your transaction, plus every other account holder **j**. This will change the transaction set to $T_{ji}= \{YN_{ji}, DT_{ji}, L_{ji}, PP_{ji}, Pr_{ji}, MP_{ji}, S_{ji}\}$ where $i=\mathbf{1},...N_j$: transaction number i, for account holder j. How can we model this with TML? One way to model this is to create individual data streams for each field type F and account holder (j). For T_{ji}, we can create seven streams for each record type, {YN, DT, L, PP, Pr, MP, S}, containing N_j amount of data in each stream and M number of streams of each type. Let's assume N=100 and M=1000, then you will have 7000 (=F x M) data streams. For each data stream type {YN, DT, L, PP, Pr, MP, S}, for every account holder, I use **viperanomalytrain** to find 7000 peer groups (P_j) of normal or usual entries, say, $P_j=\{YN_{jp}, DT_{jp}, L_{jp}, PP_{jp}, Pr_{jp}, MP_{jp}, S_{jp}\}$, where p=1...Z and Z is a large number which represents different peer groups for each record type R. Now, as I get new transactions from account holders (j), I can compare each transaction record type against its peer group, $\{YN_{jp}, DT_{jp}, L_{jp}, PP_{jp}, Pr_{jp}, MP_{jp}, S_{jp}\}$, and predict if this new transaction value is "abnormal." To do the anomaly prediction, I will use the second function **viperanomalypredict**:

```
viperanomalypredict(vipertoken,host,port,consumefrom, produceto,consume
inputstream,produceinputstreamtest,produceridinputstreamtest,
streamstoanalyse, consumeridinputstream, companyname, consumerid,
producerid,flags, hpdehost, viperconfigfile, enabletls=1,partition=-1,
hpdeport=-999,brokerhost='', brokerport=9092,delay=1000, timeout=120,
microserviceid='')
```

The preceding function is described in Table 6-6.

Table 6-6. *Viperanomalypredict Description*

Variable	Description
Vipertoken	VIPER token.
Host	Host IP address VIPER is listening on.
Port	Port number VIPER is listening on.
Consumefrom	Kafka topic containing the peer groups.
Produceto	Kafka topic to produce the results of the anomaly prediction.
Consumeinputstream	Joined topic of input streams of new transactions. Get from ***viperproducetotopicstream***.
Produceinputstreamtest	Topic name of formatted input stream to test for anomalies.
Produceridinputstreamtest	Producer ID of the test input stream.
Streamstoanalyse	Streams to analyze – same as in ***viperanomalytrain***.
Consumeridinputstream	Consumer ID for the input stream.
Companyname	Company name.
consumerid	Consumer ID for the **consumefrom** topic.
producerid	Producer ID for the **produceto** topic.
flags	You can define flags for each of the variables specified in the **Streamstoanalyse** variable. The flags must have the following format for **numeric** streams: [**riskscore**=this is the threshold for the score and must be between 0 and 1. Any computed value equal to, or above, this number is considered to be an anomaly]~[**complete**=(and, or, p[0-100] – if complete=and, then all computed risk scores must exceed the threshold risk score; if complete=or, then at least one stream must exceed the risk score for transactions to be flagged as an anomaly; if complete=p50, then at least 50% of the fields must contain anomalies)]~[type= (and,or),(topic name),(topic type),(v1=some value you want to flag as suspicious – this is optional),(sc=standardized score threshold that flags a transaction as suspicious – this is a number and ranges between 0 and 1; high values are suspicious)].

(continued)

Table 6-6. (*continued*)

Variable	Description
	For text streams the flags are [type=(or,and),(topic name), (topictype=string), (stringcontains=1 or 0; if 1, flag new transactions that contain string in v2; otherwise it will equate the string with v2),(v2=string value you want to flag as suspicious; you can use valueany or a string value. You can also use ^ (and) and | (or) to separate multiple strings),(sc=score threshold value for the stream, usually between 0 and 1)]
Hpdehost	HPDE host IP address.
Viperconfigfile	Full path of the VIPER.ENV file.
enabletls	If 1, then Kafka is enabled for SSL/TLS encryption, otherwise not.
Partition	Partition number containing the peer group to test against. This partition number is returned from ***viperanomalytrain***.
Hpdeport	HPDE port number.
Brokerhost, brokerport, delay, timeout, microserviceid	Same description as before.

Further details on how the flags determine the risk score are as follows. Given the peer group $P_j=\{YN_{jp}, DT_{jp}, L_{jp}, PP_{jp}, Pr_{jp}, MP_{jp}, S_{jp}\}$ for each account holder *j*, and given new transaction *k*: $NT_{jk}=\{YN_{jk}, DT_{jk}, L_{jk}, PP_{jk}, Pr_{jk}, MP_{jk}, S_{jk}\}$, an analysis is done to compare each field, *k*, against its peer group *p*, for every account holder *j*. To determine the computed risk score, if **complete=and**, then transaction *k* needs to have every field to be flagged as anomalous.[32] If **complete=or**, then at least one of the fields needs to be flagged as anomalous; if **complete=p50**, then at least 50% of the fields need to be flagged as anomalous. For example, if the peer groups have the following values:

1) YN={Sushma, Sushma, Sushma, Sushma, Sushma, Sushma,}

2) DT={2020-01-30-13:00, 2020-02-03-12:12, 2020-02-10-15:30,2020-03-10-10:12,2020-03-13-09:10,2020-04-13-17:10}

3) L={Toronto, Toronto, Toronto, Toronto, Toronto, Toronto}

---

[32]Fields are treated independent of other fields.

4) PP={USB stick, Hard Drive, Computer Memory, Laptop, Monitor, Book}

5) Pr={50,60,90, 50, 650,900}

6) MP={Credit Card, Credit Card, Credit Card, Credit Card, Credit Card, Credit Card, Credit Card}

7) S={Walmart, Walmart, Walmart, Walmart, Walmart, BestBuy}

And a new transaction, *k,* comes in:

1) YN=Sushma

2) DT=2020-06-30-20:00

3) L=Mexico

4) PP=Diamond Ring

5) Pr=25000

6) MP=Credit Card

7) S=Cartier

The algorithm will likely flag this because the price of 25,000 is a lot different than the peer group prices (Pr). As well, the location (L), Mexico, is different from the normal location of Toronto. And the store location (S) is much different than Walmart. This could be a case when Sushma's credit card is stolen and someone is making large purchases from her credit card. While this is a trivial example, when you are faced with many fields, transactions, and account holders, you can see how quickly this problem can grow. If you have 50 fields, 150,000 transactions per account holder per year, and 1,000,000 account holders, that is 50 million data streams, with 150,000 streaming transactions in each account. Conventional machine learning processes cannot handle this amount of scale and speed, but TML can make it relatively easy to manage with just two functions: ***viperanomalytrain*** and ***viperanomalypredict***. Specifically, using the ***viperanomalytrain***, we create the peer groups using a sliding window of transactions in real time for each account. This peer group is then used by ***viperanomalypredict*** to make the predictions of whether a new transaction is an anomaly or not for every account. These outcomes are visualized in real time, allowing you to see which transactions should be investigated for fraud. The next section discusses the bank fraud example.

Anomaly Detection on Banking Transactions with TML

This book also provides example code (set 2) to detect anomalies on simulated bank transactions.[33] This is performed using unsupervised learning because, as mentioned, it is impossible to classify the dependent variable for fraud or no fraud in real time as prior knowledge of fraud is needed to classify the dependent variable: with real-time data, prior knowledge is not readily available. TML has a unique and powerful anomaly detection process that can also be used on many other anomaly use cases such as predicting asset failures in IoT devices, predicting product recommendations, and other use cases which require a risk prediction in the form of a probability. This process is shown in Figure 6-10.

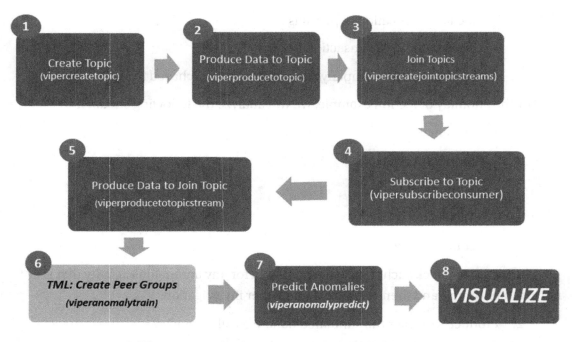

Figure 6-10. *TML Flow Diagram for Unsupervised Learning*

---

[33]All program files are on GitHub.

TML for anomaly detection follows a similar process for supervised learning, with minor changes as shown in Figure 6-10. It is useful to walk through the entire process for this example; we will highlight the core pieces of code in the bank fraud detection program.[34]

1. Create Topics creates topics for 50 simulated bank accounts with 7 fields in each account. This will result in 350 data streams in Kafka, each with 1 partition. In the example, each account is numbered from 0 to 50, and the fields are

   - transactiondatetime – Date and time of the transaction

   - currency – Currency used to buy the product or service

   - productpurchased – Product purchased

   - amountpaid – Amount paid for the product or service

   - location – Location of purchase

   - transactionid – Transaction id

   - counterparty – Company the product was purchased from

For the anomaly detection example, we will analyze the following streams:

   - currency

   - productpurchased

   - amountpaid

   - location

   - counterparty

The idea is to check each transaction and field for any anomalous activity that is not normal (abnormal) and then visualize it for further investigation.

2. Produce to Topics the simulated data. We will leverage parallel processing in Python to produce data to each stream concurrently.[35]

3. Subscribe to Topics is done to process the data.

---

[34]https://github.com/smaurice101/Predict-Bank-Fraud
[35]The program file that produces data is called Produce-Bank-Fraud-Data-to-Kafka.

4. Join the Topics creates a template topic for the joined topics (data streams): currency, productpurchased, amountpaid, location, counterparty.

5. Produce Data to Join Topics once the joined template is created. It will roll back the data streams to create historical data.

6. Create the training dataset and perform TML for anomaly detection. You create training datasets and perform TML using the function viperanomalytrain. This function will read the consolidated data and convert the JSON to an array for peer group analysis. Peer group analysis is a unique unsupervised learning algorithm that will analyze all data streams and generate a group of normal values. New transactions will be compared to these peer groups of normal values, and if the new values deviate from the normal values, they are considered non-normal or an anomaly. You can specify flags that control how sensitive HPDE is in constructing the peer group. Details on how to use flags can be found on the TML Python page.[36]

    -   Code example:

    -   *result*=maadstml.viperanomalytrain(VIPERTOKEN,VIPERHOST, VIPERPORT,*consumefrom*,*produceto*, *producepeergroupto*, produceridpeergroup,consumeridproduceto, *streamstoanalyse*, companyname,consumerid,producerid,*flagstraining*,hpdehost, viperconfigfile, enabletls,*partition*,hpdeport)

        -   *consumefrom* – Topic that contains the joined data.

        -   *produceto* – This is an intermediary topic that prepares the data for peer group analysis.

        -   *producepeergroupto* – This is the topic that will store the peer group.

        -   *streamstoanalyse* – These are the streams to analyze. You can join any number of streams, and they will be analyzed.

---

[36]https://pypi.org/project/maadstml/

- *flagstraining* – The flags you can set to control the sensitivity of how the peer groups are chosen.

- *partition* – This is the partition that contains the data to be used for peer group analysis.

- *result* – This will indicate the partition that contains the peer groups.

7. Predict Anomalies using the peer groups. Use the function viperanomalypredict to grab the new transactions and compare each new transaction against its peer group. You can specify flags that control how sensitive HPDE is in predicting anomalies. Details on how to use flags can be found on the TML Python page.[37]

- Code example:

  - result=maadstml.viperanomalypredict(VIPERTOKEN, VIPERHOST, VIPERPORT,*consumefrom*,*produceto*, *consumeinputstream*, *produceinputstreamtest*, produceridinputstreamtest, streamstoanalyse, consumeridinputstream, companyname, consumeridmainpredict, producerid, *flagsprediction*, hpdeh ost,viperconfigfile,enabletls,*peergroup_partition*,hpdeport)

  - *consumefrom* – This is the topic that contains the peer group. This is the same topic in producepeergroupto.

  - *peergroup_partition* – This is the partition that contains the peer group.

  - *produceto* – This is the topic that will contain the anomaly predictions. *You use this topic in the visualization.*

  - *consumeinputstream* – This is an input stream of new transactions that are used for anomaly predictions.

- *produceinputstreamtest* – This is the formatted data of the new transactions.

- *flagsprediction* – You can specify flags to control the sensitivity of the predictions.

8. Visualize the anomaly results and decide if a transaction needs further investigation.

The next section concludes this chapter.

Concluding Remarks

This chapter has discussed and shown how TML solutions can be built using a template that can be repurposed for other solutions. There are four key components to a TML solution:

1) Apache Kafka

   a. We used Confluent Cloud to create a Kafka cluster.

   b. We retrieved the Kafka bootstrap server address and updated the MAADS-VIPER environment file: VIPER.ENV.

   c. We retrieved the Kafka Cloud Key and Secret as our cloud username and password and updated the VIPER.ENV file.

   d. We enabled SSL/TLS encryption.

   e. We converted the keystore files into PEM formats.

2) MAADS-VIPER

   a. We downloaded MAADS-VIPER and assigned it a host and port.

   b. We got a valid Vipertoken that is required to run VIPER.

   c. We used the Windows version of VIPER.

3) MAADS-HPDE

   a. We downloaded HPDE and assigned it a host and port.

   b. We got a valid HPDE token that is required to run HPDE.

    c. We installed HPDE on the same network as VIPER.

    d. VIPER communicates with HPDE over the TCP/IP network.

4) MAADS-VIPERviz (discussed in the next chapter)

    a. We will visualize the prediction, anomalies, and optimization results in real time using VIPERviz.

5) MAADSTML Python library

    a. We installed the MAADSTML Python library using

        i. pip install maadstml

    b. We used the VIPER Python functions using the MAADSTML Python library for TML.

    c. We used this library to build the TML solution template.

This template showed the basic way to build TML solutions using just 11 core functions:

1. vipercreatetopic

2. viperproducetotopic

3. vipersubscribeconsumer

4. vipercreatejointopicstreams

5. viperproducetotopicstream

6. vipercreatetrainingdata

7. viperhpdetraining

8. viperhpdepredict

9. viperhpdeoptimize

10. viperanomalytrain

11. viperanomalypredict

We did advanced auto machine learning using HPDE with data streams in less than an hour. Using different combinations of the preceding functions, you can create powerful, frictionless, and elastic TML solutions with big data in the cloud with Apache Kafka, using Confluent Cloud running on Google Cloud Platform (or any other cloud vendor such as Amazon or Microsoft). This approach also makes it easy for a wide range of developers from beginner to advanced to start creating TML solutions from data streams with Python. The possibilities are almost endless on the types of solutions that can be created.

We discussed the TML solution flow diagram. Every TML solution will use some combination of the preceding functions. Subscribing to topics to get the consumerid is required to consume from topics. Several consumers can consume insights from the same topic using VIPERviz. VIPER can create consumer groups in Kafka. A consumer group allows multiple consumers to consume from the same topic in parallel: this is how we can do parallel processing in Kafka. When creating consumer groups, it is important to ensure the number of partitions for the topic, which the group will consume from, equal the number of consumers. So, if 100 people (consumers) will consume from the topic, then the topic should have 100 partitions. In the topics we created, we used one partition; you can easily increase this number.

Training, anomaly detection, predictions, optimization, and visualization are core components of a TML solution. HPDE is used to perform

- Model training

- Anomaly detection

- Prediction

- Optimization

VIPER visualization is used to visualize the anomalies, predictions, and optimizations. The process of training is dependent on the training datasets that are stored in Kafka topics. One caveat: Training datasets should not be large because Kafka will not store large datasets in a topic to maintain data integrity; usually, data lengths of 100 should be fine, but you need to experiment with this. Developers will need to experiment with the rollback feature (i.e., maxrows feature in viperproducetotopicstream) when producing to topic streams to ensure Kafka will accept the amount of data. TML is meant for transactional data with frequent learning from transactional data. This means there should be little need for large training datasets because learning is more frequent, as opposed to CML, where learning is not very frequent, thus requiring larger datasets. Optimization is also an important component for prescriptive analytics. VIPER has a unique way of finding global maxima or minima from trained optimal algorithms. This allows users to quickly find values for the independent variables that minimize or maximize the dependent variable.

Predictions are another important component of TML solutions. Aside from optimization, predictions are why we do supervised machine learning in the first place. This is in contrast to anomaly detection that requires unsupervised machine learning. The TML technologies are flexible to handle both supervised and unsupervised learning solutions.

There are a few areas of focus when building TML solutions:

1) Cloud cost

2) Model management – Number of algorithms

3) Security

The cloud costs should be closely tracked especially if you are creating many Kafka topics with many partitions per topic. The bigger the TML solution, the more closely costs should be tracked. AiMS offers many ways to have better visibility on Egress and Ingress of TML solutions, as well as giving administrators the ability to control the execution of TML solutions by activating or deactivating topics quickly, which leads to elastic TML solutions.

While AutoML is a great step forward in building frictionless TML solutions, this also increases the amount of algorithms that are created. This complicates model management in three areas:

1) Quantity of active models

2) Access to models

3) Servicing models if and when they stop working

The preceding three areas can quickly make management of large, multiple, TML solutions harder if not given proper visibility. Again, AiMS makes this easier for administrators by using automated alerts and notifications for topics. Servicing TML solutions that are made up of several different topics can get complicated very quickly. But this complication can be minimized by indicating for each topic:

1) Company name

2) Contact name

3) Contact email

4) Location

5) Description of topic

When topics are created, the preceding information needs to be entered for the topic. In case something happens to the TML solutions, the administrator can refer back to the preceding details and make people aware of the issues so they can be quickly addressed.

Reduced friction and high elasticity of TML solutions make them ideal for organizations that want to build highly advanced machine learning solutions that can pivot quickly with the changing needs of the business. Reducing human touchpoints has several advantages:

1) Faster development and deployment of machine learning solutions

2) Faster realization of insights from the solutions

3) Faster decision-making and therefore faster realization of business value

Increased elasticity of TML solutions has the following advantages:

1) Better control of costs from TML solutions such as cloud compute, storage, partitions of topics, and throughput

2) Better use of business resources by controlling, activating, or deactivating topics or solutions that are no longer used or needed

3) Scaling TML solutions up or down given the need for the insights from the solution

The preceding characteristics are unique to TML solutions and differentiate them from CML solutions. However, while there are many use cases for TML, there are also many for CML. The choice between TML and CML approaches depends on the type of use cases that utilize fast data, with a need for fast decision-making. As we claim, fast data will require fast machine learning. The purpose for fast machine learning should be aligned with the business needs and the business decisions that flow from fast learnings. TML solutions should not be pursued if there is no need to make fast decisions from fast data.

This chapter discussed a TML solution template that was applied to two use cases to predict Walmart foot traffic and bank fraud detection with example code. TML solutions can be built with the MAADSTML Python library. This library has all the functions to connect to VIPER and HPDE to perform machine learning for supervised and unsupervised learning from data streams. The template requires beginner knowledge of Python, and only a handful of functions are needed to build powerful TML solutions. This aligns with the TML principle of low code.

The template can also be part of larger solutions that integrate TML with CML solutions. For example, output from algorithms that are built using CML can be used in conjunction with output from TML solutions. If detecting fraud in data, using TML with CML for fraud detection can be useful to validate fraud from both methods to reduce the chances of false positives and false negatives.

Scaling TML solutions across a distributed network of Kafka brokers makes them ideal for use cases that require individual and customized machine learning models. For example, checking every bank transaction for fraud, for every account holder, using individual unsupervised machine learning models is possible with TML. Building individual machine learning models for every car is possible. Building machine learning models for every IoT device is possible with TML and so on. The limit of TML is not just the hardware limitations, but the limitation of our creative imaginations and solution architecture. The next chapter discusses the visualization of the insights from TML solutions.

CHAPTER 7

Visualize Your TML Model Insights: Optimization, Predictions, and Anomalies

Visualization of streaming insights is a critical part of decision-making. In order to understand what TML solutions are doing, you need to visualize their outputs. Also, given the nature of data streams, these visualizations need to be in real time that are pushed to your browser over secure HTTPS connections.[1] For the two examples provided, you will want to visualize

1. Walmart foot traffic prediction output

   a. How much foot traffic is being predicted at each Walmart location, at each hour, and for each month?

      i. Decisions you can make – Forecast hiring changes, inventory amounts, product promotions, and so on

   b. What the optimal time of day, month, and store location generates the highest foot traffic?

      i. Decisions you can make – Best or worst time to have more or less employees working, when to do store maintenance, when to stock shelves, and so on

---

[1]You can also visualize over HTTP.

© Sebastian Maurice 2021

S. Maurice, *Transactional Machine Learning with Data Streams and AutoML*,
https://doi.org/10.1007/978-1-4842-7023-3_7

2. Bank fraud prediction output

    a. Which bank transaction is fraudulent?

        i. Decisions you can make – Which bank transaction to investigate, which account to freeze immediately, and so on

To perform streaming visualization, you will use VIPERviz technology to visualize streaming insights. Specifically, HPDE will apply machine learning to streaming data for

1) Predictive analytics

    a. You can visualize these results with VIPERviz by pointing your browser to **Prediction.HTML**.

2) Prescriptive analytics or optimization

    a. You can visualize these results with VIPERviz by pointing your browser to **Optimization.HTML**.

3) Anomaly detection using unsupervised learning

    a. You can visualize these results with VIPERviz by pointing your browser to **Anomaly.HTML**.

4) Generic visualization

    a. You can combine topics generated by predictions, optimization, anomaly, and raw data streams and visualize with VIPERviz by pointing your browser to **Generictopics.HTML**.

5) AiMS Dashboard for algorithm management can be visualized by pointing your browser to **Aims.html**.

The HTML files are provided to you for download.[2] They should be copied to a folder **/viperviz/views** in the same location you installed VIPER.

VIPERviz uses WebSockets to connect to web browsers over secure HTTPS connections.[3] Users access the visualization using standard browsers to connect to VIPERviz. The format of the URL must be the following for each type:

---

[2]https://github.com/smaurice101/MAADS-VIPERviz
[3]It can also connect over HTTP.

1. Prediction results from the Walmart example

   a. `https://127.0.0.1:8003/prediction.html?topic=otics-tmlbook-walmartretail-foottraffic-prediction-results-output&offset=-1&groupid=&rollbackoffset=10&topictype=prediction&append=0&secure=1&consumerid=[Enter Consumer ID for Topic=otics-tmlbook-walmartretail-foottraffic-prediction-results-output]&vipertoken=hivmg1TMR1zS1ZHVqF4s83Zq1rDtsZKh9pEULHnLROBXPlaPEMZBEAyC7TYO`

   b. URL fields

      i. `https://127.0.0.1:8003/` – This assumes you started VIPERviz on port 8002 (HTTP); port 8003 will automatically be created for you and will serve HTTPS connections.

      ii. **prediction.html** – Use this file.

      iii. **topic=*otics-tmlbook-walmartretail-foottraffic-prediction-results-output***

         1. For the Walmart example, the output topic is ***otics-tmlbook-walmartretail-foottraffic-prediction-results-output***; predictions are stored in this topic.

         2. **offset**=-1 – Means go to the latest prediction results in the stream.

         3. **groupid**= – If you created a consumer group, you can put the name here. Consumer groups can be created with the MAADSTML Python library or in AiMS.

         4. **rollbackoffset**=10 – Means roll back the stream by 10 offsets and show the last 10 latest predictions.

         5. **topictype**=prediction

         6. **append**=0 – Means do *not* append data to the browser. If append=1, then data will accumulate in your browser. This could be problematic as you will accumulate MBs of data which will likely crash your browser.

       7.   **secure**=1 – Means to use HTTPS (port: 8003); if secure=0, then HTTP (port: 8002) will be used.

       8.   **consumerid**=[Enter Consumer ID for Topic=otics-tmlbook-walmartretail-foottraffic-prediction-results-output] – This consumer id will be outputted from the Walmart foot traffic program file.

       9.   **vipertoken**=hivmg1TMR1zS1ZHVqF4s83Zq1rDtsZKh9pEULHnLR0BXPlaPEMZBEAyC7TY0

2.   Optimization results from the Walmart example

   a.
```
https://127.0.0.1:8003/optimization.html?topic=otics-
tmlbook-walmartretail-foottraffic-optimization-
results-output&offset=-1&groupid=&rollbackoffset=10&
topictype=optimization&secure=1&append=0&consume
rid=[Enter Consumer ID for Topic=otics-tmlbook-
walmartretail-foottraffic-prediction-results-output]&
vipertoken=hivmg1TMR1zS1ZHVqF4s83Zq1rDtsZKh9pEULHnLR0
BXPlaPEMZBEAyC7TY0
```

   b.   URL fields

       i.   All fields are similar to the prediction URL with the following changes:

          1.   **optimization.html**

          2.   **topic=otics-tmlbook-walmartretail-foottraffic-optimization-results-output**

          3.   **topictype=optimization**

          4.   **consumerid=[Enter Consumer ID for Topic=otics-tmlbook-walmartretail-foottraffic-prediction-results-output]** – This will be outputted for you in the Walmart prediction and optimization program.

3. Bank fraud anomaly detection

   a. `https://127.0.0.1:8003/anomaly.html?topic=otics-tmlbook-anomalydataresults&offset=-1&rollbackoffset=10&append=0&topictype=anomaly&secure=1&groupid=&consumerid=[Enter your Consumer ID]&vipertoken=hivmg1TMR1zS1ZHVqF4s83Zq1rDtsZKh9pEULHnLROBXPlaPEMZBEAyC7TYO`

      i. URL fields

         1. All fields are similar to the prediction URL with the following changes:

            a. **anomaly.html**

            b. **topic=otics-tmlbook-anomalydataresults**

            c. **topictype=anomaly**

            d. **consumerid=[Enter your Consumer ID for otics-tmlbook-anomalydataresults]** – This will be outputted in the bank fraud program.

4. Generic visualization

   a. You can combine the data streams and visualize them.

   b. For the Walmart use case, you can use the following URL:
   `https://127.0.0.1:8003/generictopics.html?topic=otics-tmlbook-walmartretail-foottraffic-prediction-results-output,otics-tmlbook-walmartretail-foottraffic-optimization-results-output&offset=-1&groupid=&rollbackoffset=1&topictype=generic&append=0&secure=1&consumerid=&vipertoken=hivmg1TMR1zS1ZHVqF4s83Zq1rDtsZKh9pEULHnLROBXPlaPEMZBEAyC7TYO`

      i. URL fields

         1. **generictopics.html**

2. **topic=otics-tmlbook-walmartretail-foottraffic-prediction-results-output,otics-tmlbook-walmartretail-foottraffic-optimization-results-output** – Combine multiple topics and separate with a comma.

5. AiMS Dashboard

   a. `https://127.0.0.1:8003/aims.html?secure=1&vipertoken=`
      `hivmg1TMR1zS1ZHVqF4s83Zq1rDtsZKh9pEULHnLROBXPla`
      `PEMZBEAyC7TYO`

   b. Use **Aims.html**

The following sections will describe each visualization file.

Streaming Anomaly Detection Visualization

Anomaly detection was discussed in detail in Chapter 6. Part of successfully developing an anomaly detection model is to choose the flags for the peer group and predictions. Because data streams are varied, the parameters in the flags should be chosen carefully to properly represent a reasonable peer group to minimize false positives and false negatives. VIPERviz offers powerful streaming visualization capabilities, integrated with Kafka, to push anomalies to your web browser. You can use this visualization to help fine-tune your flags for anomaly training and predictions. VIPERviz will start to stream the anomaly detection results connecting through the file anomaly.html that produces an annotated timeline chart as shown in Figure 7-1.

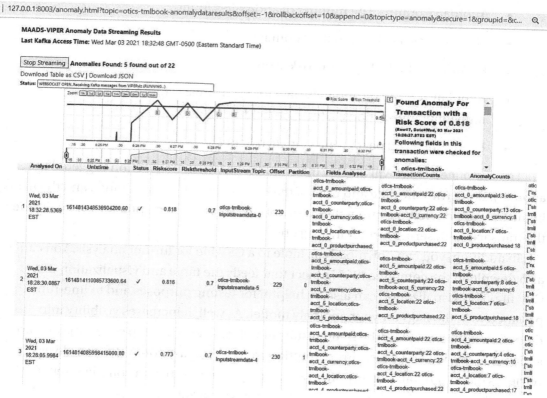

Figure 7-1. *VIPERviz Anomaly Results*

For the preceding anomaly visualization table, the following fields are displayed:

- Analysed On – Date/time of the anomaly analysis.

- Status – If X, it indicates not anomalous; if check mark, then this transaction should be looked at more closely.

- Riskscore – Compute the risk score.

- Riskthreshold – User chosen risk threshold.

- InputStream Topic – The input stream of new transactions.

- Offset – The offset of the risk score.

- Partition – The partition of the risk score.

- Fields Analysed – The data stream fields analyzed.

- TransactionCounts – The number of transactions analyzed for fraud.

- AnomalyCounts – The number of anomalies found.

- Flags – Flags used to find the anomalies.

- Kafka Key – Kafka key of the risk score.

- Anomaly data – The anomaly data for each stream; these are individual transactions, *i*, for each field *F*.

- Peer Data – The peer groups for each field *F*.

The annotated chart will display information on which transaction triggered the anomaly along with the date and time of the transaction, the row number, and the fields in the transaction that the anomalies were found. This will be helpful for you to decide if the transaction deserves further investigation.

In addition, you can download the table to a CSV file for further analysis. You can also download the complete JSON object that feeds the table and visualization to a CSV file. This visualization can also be helpful for testing purposes and to fine-tune the parameters in the flags for your anomaly model. As well, it increases visibility into the data stream that is storing the fraud results in Kafka. Users can stop streaming or start streaming; this will not cause any disruptions to the visualization. VIPERviz always retrieves data from Kafka at any offset to ensure you always have a complete set of results. The next section discusses the prediction visualization.

Streaming Prediction Visualization

Visualizing streaming predictions is another critical part of TML when developing supervised learning solutions.

Similar prediction results will be shown in Figure 7-2. The graph will plot the predictions on the Y-axis and the date/time on the X-axis. The streaming results can be stopped and started. The prediction results are graphed, and tabled, in real time and pushed to your browser. The main use of the visualization is to help you make decisions, as discussed earlier in the Walmart and bank fraud use cases. They will also help you to see the data streams and how they are behaving, and this could be useful to refine your TML models during testing. The auto-generated annotated text provides additional information that will support your decision-making. You can also download the results in the table to a CSV file. The data pushed to your browser are in JSON format; the JSON data can also be downloaded.

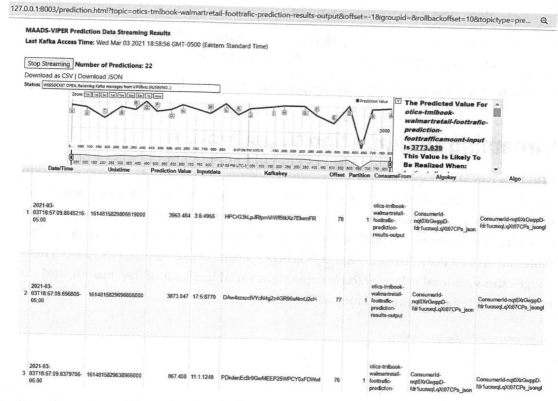

Figure 7-2. *Streaming Prediction Results*

Figure 7-2 will display the following fields:

- Date/Time – Of when the prediction was generated.

- Unixtime – Unix time of the prediction.

- Prediction Value – The prediction value.

- Inputdata – The input data (values of the independent variables) to generate the prediction.

- Kafkakey – The Kafka key for the prediction.

- Offset – Offset of the prediction.

- Partition – The partition of the prediction.

- ConsumeFrom – The topic for the trained algorithm used to generate the prediction.

- Algokey – The unique key for the algorithm generated by HPDE.

- Algo – The name of the algorithm.

- Usedeploy – If 1, then the algorithm in the /deploy folder is used; otherwise, the algorithm in the /models folder is used.

The next section discusses streaming optimization visualization.

Streaming Optimization Visualization

VIPERviz can also graph optimization results as shown in Figure 7-3. The optimization visualization will clearly display the variable that is optimized and its optimized value; it will also show the values of the independent variables that optimized the dependent variable. The optimization JSON data can be downloaded. The streaming data in the table can also be downloaded. Figure 7-3 shows results of the Walmart use case. These gauges show optimal values of the independent variables: hour of day, month, and store number that led to maximized foot traffic.

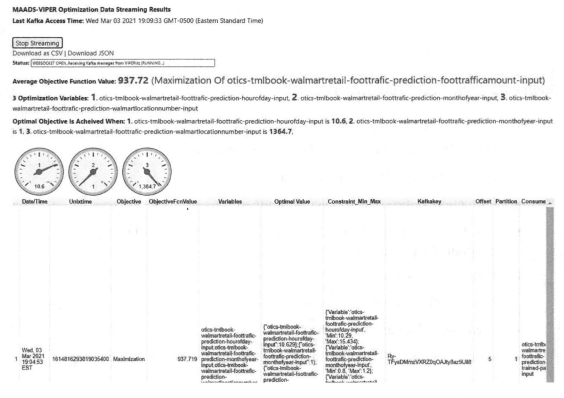

Figure 7-3. *Streaming Optimization Results*

Figure 7-3 shows the following:

- Average Objective Function Value – This is the average value of all the objective function from other optimization model runs.

- Optimization Variables – These are the names of variables that were optimized.

- Gauges show the optimal values of the independent variables.

- Date/Time – This is the date/time of the optimization execution.

- Unixtime – This is the Unix time of the optimization execution.

- ObjectiveFcnValue – The value of the optimized objective function.

- Variables – The names of variables that are optimized.

- Optimal Values – Optimal values of the independent variables.

- Constraint_Min_Max – Min and max values used for the constraints.

- Kafkakey – Kafka key of the optimization.

- Offset – Offset of the optimization.

- Partition – Partition of the optimization.

- ConsumeFrom – Topic that stores the trained parameters from training.

- Usedeploy – If 1, then the algorithm in the /deploy folder is used; otherwise, the algorithm in the /models folder is used.

The next section discusses the AiMS Dashboard.

AiMS Dashboard

AiMS is a dashboard for managing algorithms generated by VIPER and HPDE. AiMS is served by VIPERviz.

Once connected, users can control the production and consumption of every topic or group that is created by TML solutions for consumers, producers, and groups. The AiMS Dashboard is shown in Figure 7-4. AiMS was discussed in detail in Chapter 2. It is important to note that AiMS works in conjunction with VIPER. If you have several instances of VIPER running, you can use AiMS for each instance of VIPER to control

TML solutions by simply using the appropriate `Vipervizhost` and `port`. This enables developers to build large TML solutions while enabling administrators to control each solution, regardless of size, easily from the AiMS Dashboard. This modularity also makes it possible to build AiMS into a microservices architecture, with VIPER and HPDE, and decouple Kafka and AutoML processing with algorithm and model management. With this dashboard, TML solution administrators can set up alerts and notifications to determine whether topics are being used or not. If not being used, they can be deactivated. The ability to immediately deactivate TML solutions can have immediate positive impacts on costs for TML solutions, which are an important consideration especially for large cloud-based solutions.

127.0.0.1:8003/aims.html?secure=1&vipertoken=hivmg1TMR1zS1ZHVqF4s83Zq1rDtsZKh9pEULHnLR0BXPlaPEMZBEAyC7TY0

Algorithms and Insights Management System (AiMS) Dashboard

| Zookeeper: | Kafka Broker Host: Guru (192.168.0.14) | Kafka Port: 9092 | kafka Log Folder Size(MB): 8431 | kafka Active Groups: 5 | kafka Active Consumers: 10 | kafka Active Topics: 20 | kafka Total Bytes Read by Consumers(Kb): 33402854.4 | kafka Total Bytes Written by Producers(Kb): 9605893.6 |

Kafka Consumers

Notifications/Alerts:
☑ Send email when consumer(s) do not read a topic for [5] days (Enter email(s), separate multiple emails by comma) [sebastian.maurice@otics.ca] [Run]
☐ Send email every [ ] days at Email(s): [ ] when consumer(s) exceeds [ ] MB in reads from Kafka (Egress) [Run]
☐ Auto de-activate consumer(s) if they do not read a topic for [ ] days and send me an email at: [ ] [Run]
☐ Auto create a ticket in ServiceNow when consumers do not read a topic for [ ] days (Enter ServiceNow URL to POST to: [ ] [Run]
☐ Auto Generate Excel Every [ ] Days and Email Report To: [ ] [Run]
Viper Status Message: [Active]

Save Alerts | Subscribe Consumer | Modify Consumer Detail | Batch Consumer Activate | Activate | Deactivate

Search Consumers

Check All:☐	Activate/Deactivate	Topic / Algorithm	LastReadofTopic	BytesRead(kb)	ActiveReadDays	ConsumerId	ProducerId	GroupId	Company name	Contact name	Contact email
☐	De-activate	anomalypeergroup		0	0	ConsumerId-1eTptg424X5E38MmZR0tOjsKLUWtcM	ProducerId-6JJJjREfUJb8qH1j9Z4WEcnb181-dr	n/a	OTICS Advanced Analytics	Sebastian	Sebastian.Mauri
☐	De-activate	hyper-predictions2	2021-02-16	4810834.9	8	ConsumerId-3xWAD8S8HU2Gks1ArbDu3qPXR0tfi	ProducerId-CXGJHEz-RfWLx00pOo27RV1GIfb2-9	n/a	OTICS Advanced Analytics	Sebastian	Sebastian.Mauri
☐	De-activate	hyper-predictions	2021-02-16	92161	1	ConsumerId-76qV2YpL3yqXdmZDV4MsZnzTYG8i0Q	ProducerId-Ek8pe8SxD9qv8QhKTqbDyi0po2E5dE	n/a	OTICS Advanced Analytics	Sebastian	Sebastian.Mauri
☐	De-activate	joinad-viper-test15	2021-02-16	12329579	16	ConsumerId-JCMzBDqxVtRByEB7w7D01oGSe4MY57	ProducerId-yoO7Q8o8S9wcan5Bd7HHDi2P7G35Dx	n/a	OTICS Advanced Analytics	Sebastian	Sebastian.Mauri

Kafka Producer Topics/Algorithms

Notifications/Alerts:
☐ Send email when producer(s) do not write to a topic for [ ] days (Enter email(s), separate multiple emails by comma) [ ] [Run]
☐ Send email every [ ] days at Email(s): [ ] when producer(s) exceeds [ ] MB in writes to Kafka (Ingress) [Run]
☐ Auto de-activate topic(s) when NO consumers are reading from it for [ ] days and send me an email at: [ ] [Run]
☐ Auto create ticket(s) in ServiceNow when producers do not write to a topic for [ ] days (Enter ServiceNow URL to POST to: [ ] [Run]
☑ SSL/TLS is ENABLED In This Kafka Broker?
Viper Status Message: [Active]

Save Alerts | Create Topic | Modify Topic Detail | Batch Topics Activate | Activate | Deactivate

Search Producers

Check All:☐	Activate/Deactivate	Topic / Algorithm	LastWriteToTopic	BytesWritten(Kb)	ActiveWriteDays	ProducerId	Companyname	Contactname	Contactemail	Location	Descriptio
☐	De-activate	anomalydataresults	2021-02-16	120684	8	ProducerId-ajeQhiDg81qquX0dNE8fAXj9ru52qi	OTICS Advanced Analytics	Sebastian	Sebastian.Maurice	Toronto	Topic to store the anomaly results
☐	De-activate	anomalydatatest10	2021-02-16	194720.1	8	ProducerId-slXi4aoimpdPvEYs4g01IipquEhQ8hHc	OTICS Advanced Analytics	Sebastian	Sebastian.Maurice	Toronto	Topic needed for peer group

Figure 7-4. *AiMS Dashboard*

The next section discusses how to visualize multiple data streams in the same chart.

Generic Topics' Visualization

In some cases, you will want to visualize the raw data streams together. For example, in the Walmart example, you may want to visualize the independent variable streams together. Or, you may want to visualize the predictions and optimization results together; this is possible with the generic charts as shown in Figure 7-5.

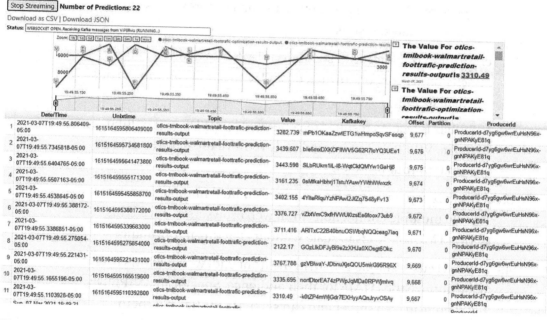

Figure 7-5. *Generic Topics Chart*

This chart is similar to the prediction chart, with the exception that multiple data streams can be visualized together. The results are also annotated to show you the value of individual streams. You can also combine text and numeric data streams.

The next section will discuss WebSockets.

Visualization with WebSockets

Using WebSockets has several key advantages for data streams:

1. Conventional browsers need to make a request to a web server to receive a response:

   a. VIPERviz does not require a web server because it has one built-in.

   b. VIPERviz does not wait for a request from a browser; it can *push* results to the web browser which ensures users always have the latest results.

   c. VIPERviz comes with standard HTML files that you can customize for your own needs. VIPERviz will continue to push data to you.

2. VIPERviz is fast. As results are generated by HPDE and written to Kafka, VIPERviz can immediately retrieve the new results and stream it to your browser.

3. VIPERviz can be instantiated to an unlimited number and can handle multiple connections from web browsers at once. This allows TML solutions to scale without affecting performance.

4. VIPERviz uses simple HTML in the front end, which allows users to customize the HTML for their audience. Specifically, VIPERviz has a built-in web server and can serve HTML/JavaScript to conventional web browsers.

5. VIPERviz is secure. It can connect using HTTPS connections with TLS encryption.

Streaming visualization over WebSockets that are secure, scalable, and portable and are integrated with VIPER/HPDE and Kafka makes TML solutions enterprise ready. The added benefit of frictionless and elasticity of TML solutions allows organizations to maximize value from data streams while allowing for mass consumption of the results by using streaming visualizations with low latency. The next section will conclude this chapter.

Concluding Remarks

TML solutions are tightly integrated with streaming visualization using VIPERviz. This allows you to build scalable end-to-end TML solutions with full visibility and transparency into what TML solutions are doing and easily use their output for decision-making. Every visualization is designed to use JSON data that is pushed to your browser by VIPERviz that is continuously monitoring Kafka topics for updates. Updates are immediately pushed to you, so no need to request it. This makes TML visualization especially effective for data streaming solutions.

You can control the data that is retrieved from Kafka using

- topic

- consumerid

- offset

- rollbackoffset

- append

- secure

- groupid

You can choose the output topics of TML solutions in the topic keyword. You can retrieve the consumer id for any topic by subscribing to the topic either through MAADSTML or through AiMS. Setting offset=-1, and a value for rollbackoffset, will give you that latest amount of data for the topic. For example, if offset=-1, and rollbackoffset=20, it will result in the last 20 data points to be sent to your browser. Also, because data streams can accumulate very quickly, you can use the append field to control how much data is appended to your browser. For example, append=0 will not append data to your browser, and you will always have the latest data even if the data in Kafka gets very large. However, if you do want to accumulate data, just set append=1. Security can also be controlled with the secure keyword. Setting secure=1 will cause VIPERviz to send data over HTTPS connections; secure=0 causes VIPERviz to switch to HTTP connections. Note that if you start VIPERviz on port 8002, then the HTTPS port is automatically the next port: 8003.

You can also visualize data from Kafka consumer groups by entering the group id in the groupid field. Kafka consumer groups are a very convenient way to scale your visualizations in a parallel fashion while maintaining maximum performance in data throughput to each consumer. You can create consumer groups programmatically with MAADSTML or AiMS.

The AiMS Dashboard will give you a complete view of what topics and groups are created and control their execution. Specifically, you can easily activate or deactivate consumer topics, producer topics, and group topics. You can also set notifications and alerts to keep track of which topics are being used and which ones are not.

Lastly, there are five, customizable, HTML files that you need to use for visualization:

1. To visualize predictions, use prediction.html.

2. To visualize anomalies, use anomaly.html.

3. To visualize optimization, use optimization.html.

4. To visualize generic topics, use generictopics.html.

5. To visualize aims, use aims.html.

All visualization data can be downloaded in CSV and JSON formats. The next chapter will discuss the future of TML.

CHAPTER 8

Evolution and Opportunities for Transactional Machine Learning in Almost Every Industry

We have discussed several areas that make TML effective in the application of AutoML to data streams for greater business insights and impact. What underlies TML opportunities is the belief that fast data will require fast machine learning for fast decision-making. Given that fast data is here to stay and will only grow in pervasiveness, it presents an opportunity for organizations to harness it and use it to add more value to their business. It is imperative that you understand the use cases and the evolution of transactional machine learning and its differences from conventional machine learning. Advantages will accrue to organizations that evaluate and build transactional machine learning capabilities earlier than their peers. To achieve this, you must automate the machine learning process together with using data streams that are real time and event-driven. Automation of the machine learning process reduces the human touchpoints and leads to a frictionless machine learning process; together with data streams, you can create solutions that are elastic and can quickly scale up or down.

There are several areas that will continue to embrace and evolve TML and are aligned with fast data and fast decision-making. These areas are

1) Faster and more complex decision-making by machines

© Sebastian Maurice 2021

S. Maurice, *Transactional Machine Learning with Data Streams and AutoML*,
https://doi.org/10.1007/978-1-4842-7023-3_8

2) Broader adoption of AutoML techniques and processes using data streams

3) Stacking or chaining TML solutions for more sophisticated use cases especially for supply chain optimization

The next section will explore each of these areas further.

Areas of Further Exploration

This section discusses each of the three areas in more detail – specifically, why these areas are of interest, what are some of the key research questions, what are some of the challenges, and the TML opportunities created for organizations.

Faster and More Complex Decision-Making by Machines

Good decisions are crucial for reducing business risk. Before machine learning, business decision-making was prone to subjectivity and slowness: analyzing hundreds or thousands of transactions for fraud, objectively by a human, is impossible. With machine learning from data, humans can complement their decision-making with data-driven insights, thereby reducing the degree of subjectivity and increasing the speed of decisions. Data-driven decision-making has several advantages:

1) It reduces the human biases in the decision-making process.

2) It makes decisions defendable.

3) It makes the decision process repeatable.

4) It increases the transparency of the decision-making process.

5) It makes the decision process auditable.

The preceding list assumes that humans are in the decision-making loop. For faster decisions, humans are a bottleneck, and in cases where speed of decision-making is important, machine learning solutions should be used to make some or all of the decisions: this is called closed-loop decision-making. Organizations will need to carefully consider the types of TML solutions they want to create and use and avoid some of the challenges that closed-loop can create. Specifically, with closed-loop decision-making:

1) How do you ensure the machine always makes the best decision?

   a. How does a machine define "best decisions"?

   b. How does a machine define "bad decisions"?

2) Should machines make life and death decisions?

   a. For example, in autonomous vehicles, how free should machines be to make decisions that could cause car accidents at high speeds?

   b. What are the controls or guard rails for machine decision-making?

3) How do we determine complex vs. noncomplex decisions?

   a. Specifically, are there decisions that are too complex for machines to make? If so, how do we determine that?

   b. How do we switch between humans and machines when decisions get too complex? What determines the point of switching?

   c. What is the role of ethics and morality in machine decision-making?

   d. How do we ensure privacy of data?

For high-speed machine decision-making, the preceding areas must also consider the speed of training a machine so it can learn from fast data. This raises a few areas of interest:

1) How can we leverage *unsupervised* learning with data streams that can quickly identify anomalies that can be further processed by supervised learning algorithms?

   a. While there are methods for unsupervised learning for data streams, there is still very little research in this area of using fast data with fast machine learning for faster fraud detection.

2) How do we ensure data quality into the decision-making process
is good?

   a. How do we define good data quality in fast data?

   b. How do we clean fast data?

3) How can statistical metrics be used in a fast data environment
to ensure the machine learning model captures the *right*
variables (data streams) for supervised learning: dependent and
independent variables?

The increasing speed of data creation will give rise to the need for increased speed
in machine learning. While faster machine learning for faster decision-making offers
many advantages for businesses, the preceding questions present areas that should be
investigated.

Broader Adoption of AutoML Techniques and Processes to Data Streams

This book has focused on several techniques and processes to show how to integrate
data streams with AutoML. It has shown how TML solutions can be built to combine and
automate:

- Data ingestion

- Data processing

- Model formulation

- Model estimation

- Model deployment

- Decision-making through streaming visualization

One of the key challenges with the preceding areas is data quality. As discussed
in Chapter 1, there are many factors that determine data quality, which could impact
the quality of the learning outcomes. Real-time data quality processes will grow in
importance as the need for faster machine learning increases in almost every industry.

Stream processing techniques have been discussed in the research communities,
mainly focused on stream mining and queries; recently, stream processing is becoming

increasingly popular commercially. Consider some of the more popular technologies from Amazon called Kinesis.[1] Kinesis will perform stream analytics using Apache Flink,[2] but it uses SQL type queries to perform analytics on data streams. This approach is no different than stream mining, not machine learning using advanced algorithms like that performed with TML. The technologies to build streaming analytics solutions have been fragmented without a clear technology stack to build solutions quickly and easily. For example, to build a streaming solution with Amazon Kinesis, you will need some or all of the following:[3]

1. AWS infrastructure

2. Kinesis

3. Apache Flink

4. Java/Scala to build the application

5. Apache Beam to analyze data

6. Apache Gelly for graph processing[4]

7. SQL

8. Other connectors

9. Third-party visualization

10. Speciality skills in these technologies

TML aims to make advanced stream processing available to a broader group of people and is democratizing advanced stream processing. Broader adoption areas are being identified as society continues to demand and generate more data from social media to autonomous/smart devices in our homes, offices, and cities. The following areas show strong promise for TML applications:

1) Agriculture – Farm machinery produces enormous amounts of fast data about maintenance, field fertilization, hyperlocal weather, and so on.

---

[1] https://aws.amazon.com/kinesis/data-analytics/
[2] https://flink.apache.org/
[3] https://aws.amazon.com/kinesis/data-analytics/
[4] For advanced analysis, users can use Apache Gelly: https://flink.apache.org/news/2015/08/24/introducing-flink-gelly.html, but this is for graph processing.

2) Health care – Medical devices and implants are connected to apps
that analyze data, such as devices for deep brain stimulation to
prevent

a. Dystonia

b. Epilepsy

c. Essential tremor

d. Obsessive-compulsive disorder

e. Parkinson's disease

3) Finance and banking – Banking consumers produce lots of
transactional data that is giving rise to FinTech and digital
currencies that can be optimized to create more segmented
products and services.

4) Retail – Retail offers many areas for fast data such as analyzing
credit card transactions for fraud, optimizing supply chains,
predicting product prices in real time, improving online and brick-
and-mortar product recommendations, and optimizing product
inventories.

5) Energy – Heavy equipment such as turbines, wind farms, solar
panels, and other field equipment produce fast data that can be
used to predict and improve asset performance, failures, and
optimization operations such as oil and gas drilling.

6) Manufacturing – Devices and manufacturing processes all
produce fast data that can be used to optimize product quality,
improve processes, and optimize supply chains.

IoT is a natural area for many applications of TML. But this area needs careful
consideration on

- The type of decisions that need to be made

- Speed of those decisions

- Location of where those decisions are made: at the edge or not

With the growing rise of 5G telecommunication networks, this will make it possible to create more data at faster speeds. Harnessing these data and applying TML shows promise, but how and where? There are many use cases, in almost every industry, that can harness TML. Specifically

- Finance and banking

  - Challenges – Improve product and service offering by removing barriers, improving security, and protecting consumers' money

  - TML opportunities

    - Lower financial losses from fraud by using TML anomaly detection

    - Segment products and services by building individual TML solutions to predict for different segments of your customer base: male vs. female, young vs. old, remote location vs. urban location, and so on

  - Important considerations

    - Data privacy and cyber-security

    - Government mandated policy and regulations

    - Scale and speed of transactions

    - Variety of devices

    - Changing consumer behaviors

- Health care

  - Challenge – Provide customized health and patient care to reduce the burden on hospitals and medical professionals and improve health and well-being for everyone

  - TML opportunity

    - Improve patient care by predicting medications and treatments using data from others that fit the demographics. For example, for a male, 50 years, suffering from joint pains, you can predict the best medication and treatment from a mobile app, consuming less doctor time.

- Predict and optimize the settings of implants based on each individual's environment, symptoms, and habits. For example, for deep brain implants, the device is implanted in a person's chest; this device is connected to a phone app, and the data can be captured and sent to a TML solution to predict and optimize the data and send it back to the phone to automatically adjust the device settings in real time.

- Build a TML solution to predict when an elderly individual may fall. Falling is the number one cause of death and injury for the elderly,[5] and there are now devices with sensors that generate data that can be used by TML solutions to predict and help alert the elderly of a potential fall.

- Important considerations

  - Data privacy and cyber-security

  - Different patient requirements and genetics

  - Scale of the number of patients

  - Risk level of patients (and doctors/nurses) to consume, and trust, digital information

  - Accuracy of machine-generated information

  - Data variety, formats, and devices

- Energy

  - Challenge – Improve energy production and increase asset life for both renewable and nonrenewable energy sources

  - TML opportunity

    - Improve life of assets by predicting their maintenance cycle within warranty periods

    - Predict when devices will fail before they fail

---

[5]www.ncoa.org/news/resources-for-reporters/get-the-facts/falls-prevention-facts/

- Predict and optimize solar and wind energy production by building TML solutions using minute-by-minute weather forecasts to make real-time predictions of load and improve pricing schemes

- Optimize oil and gas drilling by using sensor data in a well to predict and optimize control settings on a drilling rig

- Important considerations

  - Data privacy and cyber-security

  - Data and device variety

  - Data quality issues

  - Differences in data speeds and communication protocols between devices

  - Scale of the number of devices

  - Remote locations with spotty or no Internet connectivity

- Retail

  - Challenge – Provide better, and varied, products and services at competitive prices both online and on-premise while reducing supply costs

  - TML opportunity

    - Predict better and faster product recommendation for consumers

    - Improve marketing campaigns by using TML to target customers in specific segments and locations in real time

    - Improve product inventories by predicting demand for every product in real time

    - Predict the prices of every product and improve the profit margins across thousands of products at once

    - Reduce credit card fraud at the point of purchase

- Important considerations

  - Data privacy and cyber-security threats

  - Complex processes

  - Scale of credit card transactions

  - Device and data variety

  - Scale of retail customers both online and on-premise

  - Scale of the number of products and services

  - Speed of information delivery to devices and POS systems

  - Seasonality of consumer behaviors

- Agriculture

  - Challenges – Improve crop yields and production while reducing input costs and improving equipment reliability

  - TML opportunity

    - Improve crop yields by using TML to optimize the fertilization cycles for every crop using hyper local weather forecasts in real time

    - Improve the maintenance and performance of equipment by using TML to predict when they will fail, and do it within the warranty period of the equipment to reduce costs

    - Automate and optimize processes and build more closed-loop solutions that can auto-adjust device settings without human intervention

  - Important considerations

    - Data and device variability

    - Remote locations with spotty or no Internet connectivity

    - Complex decision use cases for closed-loop, and system of systems, opportunities

- Manufacturing

  - Challenge – Improving manufacturing processes, optimizing supply chains, and reducing input costs while improving product quality

  - TML opportunity

    - Connect different supply chain process and daisy chain TML solutions that feed forward intelligent information to downstream processes

    - Improve product quality by predicting product defects in real time

    - Improve asset performance by predicting end of life and improve uptime of operations

  - Important considerations

    - Data privacy and cyber-security threats

    - Device interconnectivity with different communication protocols

    - Data and device variety

    - Complex decision use cases for closed-loop, and system of systems, opportunities

    - Quality control at scale

    - Scale of the number of inputs

    - Large and complex global supply networks

Let's expand further on some of the preceding areas. The promise of smart devices, and even smart cities and countries, is slowly becoming a reality. The challenge with IoT is not the data, but device variety, which makes a stronger case for a TML solution that is device agnostic. Different varieties of devices will have different communication protocols that will make data extraction challenging, further complicating the steps to automate the operations of these devices. Health care offers many areas for TML applications. The growing cost of health care around the world driven by pandemics, diseases, and human behaviors such as lack of exercise, poor diets, and stress, and the ongoing organization and support of hospitals and clinics by governments and private firms, make health care an interesting area for TML. The growing rise in wearables that collect data from our bodies offers lots of opportunities for machine learning. These technologies allow humans to make their own decisions about their health and well-being without doctor intervention, for example:

1. Data streaming from wearables and implants – Hospitals are utilizing remote surgeries and telemedical advice with the use of machines, thereby reducing costs on travel and use of hospital resources. Search for vaccines and medical cures using data is fast growing.

2. Data streaming from IoT devices in health-care settings – Another interesting area in health care is in neurostimulation such as deep brain stimulation (DBS) to help moderate symptoms from Parkinson's disease (PD). Specifically, sensors are implanted in the human brain and chest, which generate data that can be streamed. This streaming brain activity data can be analyzed with TML to predict the onset of tremors and help prevent tremors in individual patients with PD. Analyzing brain activity with advanced machine learning is currently lacking [Lozano et al., 2019]. Using TML to build solutions for each patient, using their data, is an important application that can benefit millions of people; by providing them insights on how to adjust the devices they are using, which takes into account their individual characteristics, symptoms, and environment, it can improve the quality of life.

There are however some key risks which you need to be aware of:

1) Maintaining privacy of patient data

2) Improving skills of doctors, nurses, patients, and caregivers to use data-driven technologies

3) Trusting machines to make health decisions

As IT professionals in health care develop machine learning technologies, some areas to explore are

1) What areas in health care benefit from TML solutions?

2) How can doctors, nurses, patients, and caregivers best use the outcomes of TML solutions?

3) What are the appropriate health decisions for TML solutions?

4) What are the advantages and disadvantages of faster health decision-making?

5) What are the risks of faster health decision-making?

Finance and banking also has many potential applications of TML. As artificial intelligence quickly automates many of the human tasks such as financial trading, financial advice, portfolio optimization, and so on, opportunities for TML applications will rise. Some areas where challenges exist are

1) Reducing fraud and theft in banking

2) Increasing transparency in financial reporting

3) Better regulation that balances the needs of AI-based technologies while protecting consumers

4) Better tracking and monitoring of risky human behaviors in financial markets

5) Quicker auditing capabilities of financial transactions and records

Transactional banking using machine learning for fraud detection, money laundering, and other high-risk behavior monitoring will continue to gain traction with more online and automated banking activities. FinTech businesses that are leveraging more AI technologies are flourishing, making banking accessible by removing middle layers that force consumers to go through a bank. Now, we just use our phones for many banking tasks. This raises some further questions that are common in the industry:

- What are the risks and pitfalls for banking consumers?

- What protections are in place to limit inexperienced banking consumers from performing more advanced banking tasks?

In a recent case,[6] an inexperienced 20-year-old user of a FinTech was performing advanced financial trades and found himself owing close to $800,000 due to a bad trade; the stress of this potential loss resulted in him committing suicide. Had there been limits and stronger controls in the technology to alert this user before he made the trade could have prevented this fatal tragedy. TML technology could monitor, track, and predict whether a user is making a wrong financial trading decision, before they actually make it.

Another example where an online trading community drastically bid up the price of GameStop[7] (ticker: GME) stock, putting short sellers (mostly institutional investors) in a short squeeze that forced them to buy the stock while taking losses in the billions of dollars. A TML solution could have alerted other traders of a potentially anomalous trading activity happening on the stock by looking at the peer group of prices of GameStop and comparing it with the new prices which would have immediately showed that a stock that had been trading below $10 couldn't possibly, in most cases, be at $300 or $400 in a matter of hours. Speed and scale of decision-making, with machine learning models that are customized with individual users' (or stock) data, could be beneficial in these cases.

The energy industry which comprises oil and gas, electricity, natural gas, and renewable power such as wind, hydro, and solar are exciting areas for TML applications. Climate change and global warming caused by the burning of fossil fuels are gaining in importance and causing governments to propose strict limits on greenhouse gas emissions from cars, buildings, data centers, and agriculture. This is causing shifts to more environment-friendly technologies and fuel sources. An example of this shift is car manufacturers increasing their production levels of electric vehicles to meet consumers' changing demands and government regulations on climate change.

TML applications in energy are varied. Areas of interest exist in

1) Oil production and optimization

2) Retail pricing of fuel products: wind power, solar, refined gasoline, diesel

3) Drilling optimization: reducing total drilling days, increasing rate of penetration, and so on

---

[6]www.forbes.com/sites/sergeiklebnikov/2020/06/17/20-year-old-robinhood-customer-dies-by-suicide-after-seeing-a-730000-negative-balance/?sh=1e1789201638

[7]www.cnn.com/2021/01/27/investing/gamestop-reddit-stock/index.html

TML areas of exploration could address

1) How can TML solutions integrate human decisions into faster outcomes? For example, in the drilling process, humans are in control of the drilling process; how can a machine offer "real-time" recommendation to a human as they are drilling a well, at any level of depth?

2) Dynamic pricing strategies for renewables that account for weather variability across a fleet of assets are an area that could benefit from TML solutions.

3) Asset monitoring and failure prediction – With more sensors connected to field assets, the data that are generated can be used to predict failure and reduce downtime. Combining data streams from different sensors in a training dataset can be used to make complex decisions about asset performance.

Manufacturing presents many areas of interest for TML applications, especially given the Covid-19 pandemic that has strained (constrained) supply chains, making synchronization between supply chains very difficult. Areas that can help are

1) Optimizing interconnected logistics

2) Automating manufacturing processes and supply chains

3) Inventory management/invoice management

4) Optimal selection of inputs into manufactured products

5) Reducing health and safety incidents during manufacturing of products

6) Predicting defective products

Manufacturing processes can be very complex and technical. For instance, supervised and unsupervised process mining in manufacturing can identify suboptimal patterns in complex processes that drive up costs. Process mining can use both anomaly detection and supervised learning techniques to

1) Reduce the complexity in manufacturing processes

2) Identify bottlenecks in the manufacturing process to help streamline the process

3) Reduce health and safety incidents

4) Be deployed in diverse manufacturing environments with outcomes that can be consumed, by humans, quickly and easily

5) Combine information from several, potentially millions of, areas of the manufacturing process to improve the supply chain speed and cost that leads to the final product

There are other areas of exploration; hopefully this section has generated interest that can lead to other areas not mentioned. The TML application areas mentioned share a common theme of scale, speed, and machine intelligence that can lead to better decision-making by humans and machines. The challenges highlight the need for clearer planning and identification of value areas that align with corporate and societal priorities over time. The next section continues this theme to discuss integration between TML solutions for deeper, more robust, intelligence at scale.

Stacking and Chaining Different TML Solutions

TML solutions, on their own, can be designed to incorporate multiple data streams and multiple algorithms. This opens up several other areas for TML solutions as a system of systems (SoS): combining individual TML solutions to form a larger system.

The system of systems with data streams integrated with AutoML have received little attention in the research literature and by organizations, but offer opportunities to build complex autonomous systems that combine each TML solution. The ability to create new streams of data from outputs from one TML solution to be used as inputs into another TML solution creates implicit and explicit connections between solutions that can be used to build a system of systems. Quite complex TML solutions can be built using data streams. Let's take a specific example in oil and gas production optimization to show how it could use a TML SoS solution by asking the following questions:

1) How will the inputs and output be used between TML solutions?

   a. Input TML solutions would be built for each oil well that would monitor the pressure, gas, and water production (among other inputs).

   b. Output TML solution would use the inputs to adjust controls of pressure, gas, and water pressure and to determine the optimal combination of inputs to extract more oil at a faster rate.

2) What types of algorithms will be required by each solution?

a. Since each input TML solution is monitoring pressure, gas, and water, it would use individually generated AutoML algorithms specific to the data it is using. This is an important aspect of TML solutions: having the ability to create customized and highly data-specific algorithms at scale and to do it quickly.

3) When should the information be exchanged between TML solutions? Do they need to be synchronized?

a. The input TML solutions send the optimal values of the pressure, gas, and water, at specific time intervals, to the output TML solution so it can quickly adjust the control settings across the field in sub-second time.

4) How many components (TML solutions) does a system of systems require?

a. This is dependent on the size of the field and the number of oil wells; the number of devices such as pumps, bits, and drill pipes; and other machinery that is generating fast data and directly or indirectly involved in oil production.

5) Is there a controlling, orchestration, mechanism that controls each TML solution and hence the entire system? This would be similar to a master-slave type models.

a. The output TML solution would make a decision based on preconfigured business rules to automatically adjust control settings (closed-loop) or present the information to a human as decision support (open-loop).

While developing a system of systems with TML solutions may seem complicated, architecting a system of systems that answers the preceding questions will ensure the overall system meets your design objectives and can be functional in a real-world setting. It could be a powerful way to improve operational efficiencies for large and complex operations. The next section concludes this chapter.

Concluding Remarks

This chapter has provided several areas for further exploration and discussed the evolution of TML solutions. It has discussed some research questions that may provide interesting areas to explore across several industries such as energy, health care, manufacturing, retail, finance and banking, and IoT. These areas present many use cases and many challenges that require more research, but offer many opportunities for TML solutions. The TML LinkedIn page[8] is a great place to see the latest news on TML as well as the GitHub page.[9]

The issues of data privacy, data quality, faster decision-making by humans or machines, risk, supply chain optimizations, and asset performance and monitoring have real consequences and value in today's digital world. The speed of data creation, together with the speed of machine learning, is a natural evolution from the way we do machine learning today.

As humans and machines produce more data faster, the race to keep up and provide faster insights for better decision-making will be paramount to improve quality of life, generate revenue, and increase customer satisfaction. Determining how we consume and deliver faster, and higher quality, information will also be a key area of interest and growth. This is already evident in the streaming technologies that power social media, movies, and music – specifically, using machine learning to target advertising to users, providing instant recommendations on movies, music, or products so consumers watch more, buy more, and listen more. All this is just the beginning, as telecommunication networks like 5G start to become more pervasive, making it easier for more data to flow at faster speeds on different devices all at once.

The barriers to building powerful solutions at lower costs with scalable infrastructures are getting lower, and they exist today. TML is a way to harness the speed of data and machine intelligence, with the power of the cloud infrastructure to drive innovations much faster and at lower costs.

Autonomous technologies will get smarter, requiring more interconnections between technologies, opening up areas for SoSs that connect TML solutions. A system of systems approach with TML solutions could offer a seamless way to combine inputs

---

[8]www.linkedin.com/groups/13930079/

[9]https://github.com/smaurice101/transactionalmachinelearning

and outputs that push and pull insights between TML systems, with a master controller
that orchestrates the TML solutions to produce a complex outcome. These types of
systems are harder and likely more costly to implement with conventional machine
learning processes.

TML will continue to grow not because it should, but because of the increasing
data creation speed, driven by human demand for faster, smarter information; this will
drive organizations to deliver TML types of solutions. The barriers of cost, technology,
and infrastructure are lowering, leveling the innovation playing field by democratizing
advanced, intelligent technologies.

The future of TML looks bright, and it is following the innovation trajectory
of solutions that are faster, cheaper, smarter, and scalable. While TML can offer
considerable value to organizations, they must always be in alignment with the
corporate priorities and business needs.

CHAPTER 9

TML Project Planning Approach and Closing Thoughts

This book has introduced all aspects of TML, which applies AutoML to data streams integrated with Apache Kafka that allows developers to build TML solutions that are scalable, frictionless, and elastic in the cloud. The importance of frictionless and elasticity is a unique characteristic of TML solutions and a differentiating factor with conventional machine learning (CML). We also formally defined TML along with the five features of TML solutions such as data fluidity, joining data streams, standardization of data streams to JSON, integration of data streams with AutoML, and the ability to create TML solutions with low code. TML is based on the belief that fast data requires fast machine learning for fast decision-making. This does not mean that all fast data needs TML, because if there is no need to make fast decisions, then TML makes little sense.

CML is not ideal for data streams for several reasons:

- CML processes mainly operate on static, disk-resident data.

- CML processes require many human touchpoints for data preparation, model formulation, model estimation, hyperparameter tuning, and reporting, which increases the friction in CML solutions.

- CML solutions are slow in relearning from newer data due to point 2.

- CML solutions are not easy to, quickly, scale up or scale down when business needs or user consumption changes.

- CML solution management of models, deployment of models, and support of solutions are, in many cases, difficult.

© Sebastian Maurice 2021

S. Maurice, *Transactional Machine Learning with Data Streams and AutoML*,
https://doi.org/10.1007/978-1-4842-7023-3_9

TML differs from CML in the following ways:

- TML applies machine learning to continuously flowing, real-time or event-driven, data streams.

- TML automatically builds training datasets by combining or joining data streams on the fly; then it applies auto machine learning to these training datasets to find an algorithm that best fits the training datasets – with minimal human intervention leading to frictionless solutions.

- TML can immediately deploy solutions for consumption by humans or machines.

- TML can continuously, and automatically, build training datasets to relearn from newer data streams and immediately make the insights available for consumption.

- TML solutions can easily scale up or scale down by

  a. Adding or reducing consumers consuming the insights from TML solutions

  b. Adding or reducing producers of data streams

  c. Joining more streams to build bigger (or reducing data streams to build smaller) solutions with more (or fewer) machine learning algorithms quickly

  d. Automatically activating or deactivating TML solution components (producer and consumer topics) using the AiMS Dashboard or TML Python functions

While TML has advantages, I am not implying it is a substitute for CML. To the contrary, CML has its value for organizations that do not have fast data, nor do they need fast data. CML processes, in the data science community, are well established and widely used. There are many use cases for CML that do not require TML. The advantage of TML is when data are being generated quickly and continuously in real time or event-driven. But, even if data are fast flowing, it still may not justify TML, if there is no need for faster decision-making. Both CML and TML have their areas of applications, and class of problems, that are meant to add value to organizations, but they must be used within the right business context.

Data are the common thread between CML and TML. Both approaches need good data quality. Bad data will lead to bad machine learning insights (solutions). We discussed several areas of data quality that are pertinent to data streams, such as velocity, variety, veracity, and volume. Data streams are part of the big data phenomenon, but not all big data is streaming data. Data streams flow in real time, but they are also created by events. Events give rise to data creation that can trigger some action. For example, if you buy an item on Amazon, the pay completion event will trigger a product shipment action. It is easy to imagine how these scenarios have made our lives much simpler in some ways and complicated in others. Event streaming and data streaming are, more or less, synonymous. TML still requires clean data to provide insights that can be trusted and actioned. In addition, its success still hinges on effective data engineering and architecture practices that lead to TML solutions that are scalable and secure with quality of insights that can be used for effective decision-making. While machines will learn from both good and bad data, the learnings will reflect the quality of the data used. The next section discusses the TML technology stack.

TML Technology Stack

Aside from the designing and architecting of TML solutions, the technology needed to build functional solutions is dependent on

- Data stream storage platform that is capable of handling large amounts of data – Apache Kafka in the cloud running in GCP, AWS, or Microsoft cloud infrastructure is ideal.

- Source and sink connector that can ingest the data from source systems or devices and write them to Kafka – MAADS-VIPER

- Analyzing streaming data for insights using deep learning algorithms – MAADS-HPDE

- Visualizing the outcomes of the TML models in real time – MAADS-VIPERviz

The storage, analysis, and visualization of insights make TML solutions scalable, secure, frictionless, elastic, and ideal for real-time decision-making by either humans or machines.

In Chapters 6 and 7, we discussed the TML solution template with accompanying technologies that are used to build and visualize the insights from TML solutions, such as

- **Apache Kafka in the cloud** which can run on Amazon Cloud Kafka, Microsoft Cloud Kafka, Google Cloud Kafka, or Confluent Cloud Kafka and is our single source of data for TML.

- **MAADS-VIPER** is the official Kafka source and sink connector to ingest data from devices and produce data to, and consume data from, Kafka.

- **MAADS-HPDE** is the AutoML technology that is integrated with VIPER and Kafka and performs all, supervised and unsupervised, machine learning functions:

  a. Applies machine learning algorithms and finds the best algorithm for the training dataset

  b. Uses the algorithm for

   i. Predictive analytics

   ii. Mathematical optimization (prescriptive analytics)

   iii. Anomaly detection

  c. Produces all results to Apache Kafka

- **MAADS-VIPERviz** delivers the visualization of streaming insights or outcomes by pushing them to the clients' browser over HTTPS or HTTP connections from

  a. Predictive analytics – Pushes predictions from predictive TML solutions

  b. Optimization – Pushes optimization results from optimized TML solutions

  c. Anomaly detection – Pushes fraud detection results from anomaly detection TML solutions

  d. AiMS Dashboard – Pushes all information on TML solutions for algorithm and insight management

- **MAADSTML Python library**[1] allows users to create powerful TML applications with the TML Python library functions with just a few lines of code.

The preceding technologies simplify the integration with Apache Kafka and machine learning to perform advanced analysis with data streams. They also simplify complex tasks such as

- Ingesting data from devices and source systems and producing data to and consuming data from Kafka running on GCP, AWS, and Azure.

- Creating and managing Kafka topics while tracking the partitions and offsets of each data. By carefully managing partitions and offsets, VIPER can create training datasets by using the offset to roll back the data stream to get a sliding window of data history and then join multiple sliding windows (from each data stream) to build a training dataset for machine learning.

- The analysis process is complex, especially when dealing with multiple data streams and multiple partitions with millions of offsets, all with fast flowing data. TML technologies simplify the analysis process by (1) automatically applying advanced machine learning to the joined data streams and finding the best algorithm; (2) automatically predicting, optimizing, and detecting anomalies using new data with the latest trained model; and (3) pushing the new insights from any TML solution to a client's browser immediately.

- Build machine learning models in real time by combining data streams for both supervised and unsupervised machine learning.

  a. For supervised machine learning, we use data streams as a dependent variable stream and independent variable streams. By specifying which streams are dependent and independent variables, we can formulate a machine learning model in real time and send it to HPDE for training. Training with joined data streams is a very powerful feature of TML with Kafka. Specifically,

---

[1]If users do not want to use Python, all VIPER functions are accessible via REST API which makes it accessible by other programming languages.

we use the offset parameter to automatically roll back each stream and get a historical dataset (sliding window), which represents our training dataset. VIPER tells Kafka to store the training dataset in a topic. Kafka stores the training dataset in the specified topic, in a partition in the topic; it returns the partition number back to VIPER for future use. Next, VIPER calls HPDE and passes it the topic *and* the partition number containing the training dataset. This is powerful for the following reasons:

   i. Since Kafka chooses the partition to optimize storage of big data, it tells the calling function[2] which partition it chose; you then use this partition when calling HPDE which is guaranteed to use the correct training dataset for machine learning. This process allows us to do machine learning on very large datasets[3] using advanced auto machine learning with relative ease, on a continuous basis, with new and evolving training datasets in real time. We don't need to use any special programming languages to do machine learning on large datasets, that is, Scala.

  ii. Using the offset parameter, we can tell Kafka the exact offset location of the training dataset in the partition. This allows us to keep an audit trail of training datasets in the event someone wants to audit the TML models. We can create as many training datasets as we want, continue to add them to the topic, and precisely direct HPDE to the training dataset we want to use at the partition and offset locations in the Kafka topic.

---

[2]These functions are in the MAADSTML Python library or using RESTful API.

[3]By taking sliding windows of the data continuously, we are performing transactional machine learning regardless of how big the data gets over time.

b. For unsupervised learning, our process changes slightly. Rather than building training datasets, we build peer groups. These peer groups represent "normal" behaviors in data streams.

For anomaly detection, TML will do the following:

   i. Join the data streams (fields) of interest and find the peer groups in each stream (field) and store it in a Kafka topic and return the partition and offset.

   ii. Call HPDE and send it the topic name, partition, and offset of the peer group.

   iii. For each new transaction, and each field in the transaction, compare it to its corresponding peer group and compute the risk score.

   iv. Visualize the results in real time.

Simplifying the complex tasks of ingesting data from devices and source systems, producing and consuming data to Kafka, analyzing it, and visualizing it for decision-making is why TML and its technologies are breaking new ground in the area of fast data and fast machine learning that enable fast decision-making in real time and at scale. The next section discusses the TML project planning approach.

TML Project Planning Approach

The ability to handle both supervised and unsupervised learning with data streams and AutoML differentiates TML from not only CML solutions but stream processing technologies that focus more on rule-based and SQL-based approaches for predictive, prescriptive, and anomaly detection. TML for anomaly detection with data streams offers many use cases in banking and finance, IoT and asset performance, manufacturing, supply chain, and many more. The driving force behind both supervised and unsupervised TML solutions is the creation of training datasets and peer groups, respectively, with Kafka's capability to roll back data streams in real time using partitions to manage scale within a big data environment.

The preceding process can quickly become overwhelming and difficult to manage, especially for large TML solutions with thousands of data streams, algorithms, and insights. Many industries still struggle with managing machine learning solutions, regardless of the size of the solution, mainly due to

- Lack of alignment of solutions to corporate needs, strategy, and goals

- Difficulty in model management and operations, that is, machine learning solution breaks down or does not provide good results, causing people to stop using it

- Lack of support for the solution

- Unclear design objectives, which lead people to distrust the data that drives the models and the outcomes

- Poor solution performance, that is, the solution is too slow to produce the insights for decision-making, causing people to stop using the solution

The preceding struggles can be overcome by focusing on the following:

1. Alignment of the TML solution to corporate objectives: using the balanced scorecard approach will help.

2. Choosing the right people and skills needed to successfully build TML solutions. These will include the following:

   a. Data scientists who are well versed in CML but understand streaming analytics – We introduced the term data stream scientist which is a person who can build TML models by combining specific data streams to solve stream-based problems.

   b. Solution architect who has deep knowledge of how event streaming platforms (i.e., Apache Kafka) work with machine learning technologies to build scalable TML solutions that are frictionless and elastic with Apache Kafka in the cloud in a microservices architecture.

   c. Data engineers who understand the nuances of data streams – Event-driven and real-time data that have temporal locality (evolve over time).

d. Python developer who can use the TML Python library to ingest data from devices and source systems and stream it to Kafka as well as use the TML Python library functions to build TML solutions.

e. Visualization developer who has deep knowledge of

   i. JavaScript

   ii. JSON data format

   iii. Visualization APIs like Google Charts

f. Technical project manager who has been exposed to

   i. Python

   ii. Machine learning technologies

   iii. Kafka

   iv. Cloud technologies: AWS, GCP, Confluent, Azure

   v. Data streams

The TML team size will depend on the following factors:

1. Number of data streams used in the TML solution

2. Source systems or devices that will be used in the TML solution

3. Number of instances of

   a. Kafka brokers

   b. VIPER/HPDE/VIPERviz

4. Number of visualization charts

5. Number of consumers of the visualizations

For example, let's take a typical TML solution that has

1. 1000 data streams

2. 100 devices and source systems

3. Following instances:

    a. 1 Kafka broker

    b. 10 VIPER instances

    c. 5 HPDE instances

    d. 5 VIPERviz instances

4. 10 visualization charts

5. 50 consumers

6. 1 TML use case

Table 9-1 shows the breakdown of the resources and the effort days.[4]

Table 9-1. *TML Effort and Resource Breakdown*

Resource	Effort (Days)
Data (stream) scientist	14
Solution architect	11
Visualization developer	12
Python developer	28
Data engineer	20
Technical project manager	17
Total effort days	102
With 20% contingency	122

The Effort (Days) column in Table 9-1 was computed as follows:

- Data scientist – We used 4 days per use case (0.8 FTE) and 0.02 FTE[5] per source system and device, which gives us 14 days of effort. The rationale is that the data scientist will be focused on solving the use

---

[4]These effort estimates may vary for your specific use case and will likely improve (reduce) over time as familiarity with TML increases and artifacts are reused between projects.

[5]FTE=Full-Time Equivalent is someone working 5 days a week, 8 hours per day. So 1 FTE=40 hours/week.

case problem and less on machine learning tasks because TML uses AutoML. Their key responsibility will be on the formulation of the training datasets, by identifying the dependent and independent variable data streams, that will comprise a machine learning model for training.

- Solution architect – We used 0.02 FTE per source system and 0.02 FTE for the instances of VIPER, HPDE, and VIPERviz, which gives us 11 days. The rationale is that the solution architect needs to focus on how the source systems integrate with the overall TML solution. If the source systems are complex with different data formats, communication protocols, and other hardware or software complexities, this effort may change. The key responsibility for the solution architect is to design the TML architecture with the data streams and source systems that are scalable and robust that aligns with the elastic nature of TML solutions.

- The visualization developer will work (develop or configure) with the visualizations for predictions, optimizations, and anomaly detections. We used 0.04 FTE for the number of visualizations (20) and consumers (50), which gives 12 days. It should be noted that VIPERviz comes with out-of-the-box visualizations, and so they do not need to be developed from scratch. This results in considerable time and cost savings.

- The Python developer will use the MAADSTML Python library to develop the TML solution. We used 0.04 FTE per source system, 0.02 FTE for all instances of VIPER/HPDE/VIPERviz, and 7 days (1.4 FTE) per use case, which gives 28 days. Because the TML Python library has prebuilt functions, the amount of effort is considerably reduced.

- The data engineer will work closely with the developer to ensure the data is accessible and can be streamed to Kafka. There could be data cleaning and preparation work to be done. We used 0.04 FTE per source system, which gives 20 days.

- The technical project manager is an important role to keep the TML project moving forward smoothly. To compute the effort, we took 20% of the sum of effort for all the resources, which gives us 17 days.

Are the preceding effort estimates reasonable? Based on our experience, yes. If we add a 20% contingency, the TML project will take approximately 122 effort days to complete. Since many tasks can happen in parallel, the project can be done in less than one month. Figure 9-1 shows a graphical representation of resources and effort days. Note there may be other operational or project roles for support and maintenance of TML solutions, documentation, training, change management and control, and security. However, we chose not to add them and focus on the key roles needed to develop TML solutions in an effort to keep the example simple.

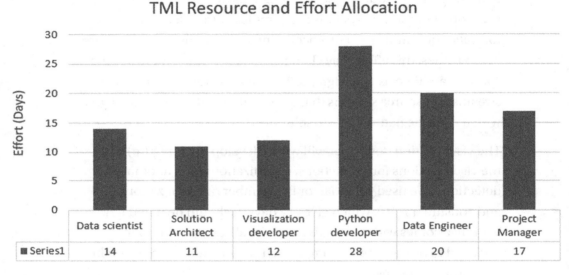

TML Resource and Effort Allocation

	Data scientist	Solution Architect	Visualization developer	Python developer	Data Engineer	Project Manager
■ Series1	14	11	12	28	20	17

Figure 9-1. *TML Resource and Effort Allocation*

The next section discusses TML value creation.

TML Value Creation

Creating value is important for every business. We also presented in Chapter 4 a TML value framework (TVF) that will help you to choose TML use cases that are prioritized and closely aligned to your corporate goals and strategy. Alignment of machine learning solutions to corporate goals and strategy ensures these solutions produce a positive profit margin: the cost of developing the solution is less than the value generated from the solution over the life of the solution. Many times businesses lose sight of the cost of advanced technology solutions before it's too late. TML, together with AiMS, offers administrators the capability to closely track and monitor solutions for (consumer) use. Because TML technologies are integrated with Apache Kafka, which uses a publisher-subscriber framework, the ability to track consumers and producers is possible. Setting alerts and notifications when no one is consuming insights from TML solutions can be important to trigger a deactivation of solutions when a period of non-use is exceeded. Especially for large solutions that are consuming cloud compute, storage, and network resources, having the ability to deactivate solutions that no one is using can result in considerable savings for organizations. These savings can be transferred to build higher-value TML solutions. A value from TML solutions is realized from

- Faster decision-making – By reducing the friction that exists in CML, TML increases speed to insights with little human intervention. For example, a TML IoT solution can continuously train itself on IoT data and predict the risk of a device failure which could alert a human who may decide to take preemptive maintenance measures.

- Faster scale – Together with Kafka, TML solutions are elastic and can scale fast using an API and microservices architecture, opening up new opportunities for organizations in almost every industry. In an IoT context, for example, hundreds of new devices can be added to a TML platform quickly with very little incremental effort on the part of the data and data science teams.

- Faster (deeper) insights – Fraud detection or product recommendations can not only be served instantly but – unlike CML solutions – TML is constantly self-optimizing and self-learning with new data, incorporating novel fraud attempts or product purchase trends with little human intervention, while delivering the latest insights in the visualizations to the client's browser for better, and easier, decision-making.

The ability to create value through better and faster decision-making will require the ability to visualize the outcomes from TML solutions. Visualization is important with data streams due to the speed of data, and insight, creation. Increases in the speed of data creation lead to a faster accumulation of data, which leads to more diverse trends and patterns in the data over time. The temporal locality of data streams gives importance to time, which will affect the learnings and insights captured. In supervised learning, the machine will capture the learning in the estimated parameters. The ability to retrain or relearn is one of the most important aspects of TML solutions. By continuously relearning, we are automatically updating the estimated parameters and ensuring we have the latest learnings from the latest data. This is not possible to do quickly with CML that is driven by static, disk-resident data. Having up-to-date estimated parameters means the machine has learned any new trends and patterns in the data. Using the latest parameters, we are more likely to make better predictions and find better optimal values for the independent variables. This is very important in industries where model degradation can happen rapidly, harming business performance and reducing customer satisfaction.

For unsupervised learning with data streams, the machine applies algorithms to the joined data stream to find "normal" behaviors. The way it defines "normal" is by computing the average or centroid value of each of the data streams. Any value that deviates significantly from the average or centroid is considered "abnormal" and not included in the peer group. How normality is chosen is by computing the standard deviation of each data point from its centroid. HPDE does this very quickly.

You can also define flags that control how peer groups are chosen for each data stream. Again, this is a very powerful way to define different normal behaviors for each data stream. Regardless, if you have 100 or 100,000 data streams, or more, the process does not change. By looking for anomalies in each data stream and then combining the results to formulate a risk score, you can very quickly determine if the behavior is normal or abnormal and then visualize it and investigate the transaction further. The next section provides some closing thoughts.

Closing Thoughts

As mentioned, TML solutions have little use if they are not used for decision-making purposes, by humans or machines, that will add value to your business. In some cases, organizations fall victim to AI and ML vendors that promise on paper how their solutions can add value to their business, but in practice fall short of actually delivering on that promise. With TML, the distance between what's on paper and what is practiced is small, because the speed to proving the value of a TML solution in a proof of concept could be done in hours. This could help establish a visible link between the TML solution and tangible business value. If machine learning solutions do not add direct or indirect value to your business, then they are wasting valuable business resources. Properly monitoring and tracking TML solutions, using AiMS, for value generation is important to ensure solutions are being used and adding value to the organization over time.

The future for TML looks bright, due in part to an unprecedented amount of data being created at faster speeds in almost every industry. As costs for cloud infrastructure along with compute, storage, and network throughput all come down, along with upgrades in telecommunication networks to 5G, the need for faster machine learning will rise. We are already seeing an increase in the adoption of cloud technologies and growth in the demand for AI and machine learning skills driven by the need for more advanced data analysis with algorithms. Many organizations are much further along in their digital transformation journeys, with data and AI technologies as part of the foundation for growth. Areas of focus for many organizations will shift to data streams and AutoML, as technologies like Apache Kafka, which is used by 80% of the Fortune 100 companies,[6] continue to provide growth opportunities for TML. Faster machine learning for faster decision-making is also required by smart devices that are helping to improve the quality of our lives around the world.

In a post-Covid world, businesses will look for ways to reduce costs, with AI and machine learning being a big part of operational efficiencies that are leading to better, and cheaper, ways of providing products and services that closely meet the needs of consumers. But, TML is not the solution for all problems; it should be part of the overall innovation strategy that puts emphasis on fast machine learning, for fast decision-making, with fast data. Only through this strategic realization will organizations prioritize TML solutions to solve the right business problems that provide the right decisions to create business value fast.

As you start to get deeper into TML with data streams and AutoML using the technologies highlighted in this book, you will start to realize there are a large class of use cases that fit within the TML umbrella. It will not only create opportunities for your business but also for you, by learning the latest ways of applying advanced auto, supervised and unsupervised, machine learning to fast flowing data streams that conventional machine learning methods cannot fully do justice.

---

[6]https://kafka.apache.org/

Definitions

The following table defines terms used throughout the book.

Name	Description
AiMS	Algorithm and Insights Management System is a dashboard that manages all algorithms created by VIPER. AiMS can also provide alerts and notifications to deactivate algorithms that are not being used.
Algorithm	A logical set of mathematical instructions followed by a computer to manipulate the training dataset to learning the patterns and the underlying structure and relationships between the dependent and independent variable(s) to achieve a desired goal.
Apache Kafka	This is an event streaming platform used to store data streams. More details can be found on www.apache.org.
Auto machine learning (AutoML)	Automated machine learning that automates data prepping, model estimation, model testing, and model reporting.
Data stream	A continuous flow of data that accumulates over time.
DSSP	Data stream storage platform that stores all data, training datasets, predictions, algorithms, and optimization results.
Elastic machine learning process	Elastic machine learning is a frictionless machine learning that creates solutions that can scale up or down very quickly with any size of data or any number of data streams. By subscribing consumers to TML solution insights, or removing consumers, administrators can scale TML solutions to any number of applications or devices or reduce the number of applications or devices immediately. Solution governance and management is also an important part of an elastic machine learning process.

© Sebastian Maurice 2021

S. Maurice, *Transactional Machine Learning with Data Streams and AutoML*, https://doi.org/10.1007/978-1-4842-7023-3

Name	Description
Frictionless machine learning process	Machine learning that has very few to zero human touchpoints in the data preparation, model formulation, feature selection, model estimation, model hyperparameter fine-tuning, model reporting, and model deployment. Frictionless machine learning solutions can scale more quickly, and elastically, with any size of data than CML solutions.
Offset	A number that marks the position of a data point in the data stream.
MAADS-HPDE	Performs AutoML on transactional data. HPDE can perform supervised and unsupervised learning. It is integrated with MAADS-VIPER and Kafka. It can be instantiated to an unlimited number of instances for load shedding. HPDE can be downloaded from here: `https://github.com/smaurice101/transactionalmachinelearning`.
MAADS-VIPER	A Kafka source and sink connector for TML. It is integrated with HPDE and Kafka. It is SSL/TLS compatible and can be instantiated to an unlimited number of instances for load shedding. VIPER can be downloaded from here: `https://github.com/smaurice101/transactionalmachinelearning`.
MAADS-VIPERviz	Performs real-time visualization for all use cases: (1) predictive analytics, (2) optimization analytics, (3) anomaly detection, (4) management of algorithms using AiMS. VIPERviz is accessed by users via a standard browser using a secure HTTPS connection. It is integrated with VIPER/HPDE and Kafka. It can be instantiated to an unlimited number of instances for load shedding. VIPERviz can be downloaded from here: `https://github.com/smaurice101/transactionalmachinelearning`.
MAADS Python library	Allows users to write low-code TML solutions using MAADS Python library. All functions (shown in Table 2-3 MAADSTML Python Library Functions) connect to MAADS-VIPER. More details on the library can be found here: `https://pypi.org/project/maads/`.
Middleware software	This is software that processes and transforms data. It also interfaces with the DSSP by producing data to the DSSP and consuming data from the DSSP.

Name	Description
Model	This is a formulation comprised on a dependent variable and independent variable(s) that an algorithm is applied to.
TML solution	A TML solution is a combination of data streams (or topics in Kafka) and auto machine learning algorithms that uses low code (MAADS Python library or REST API) as a wrapper to build and bind the components into a solution.
Topic	This is the name of a data stream in Apache Kafka.
Transactional Machine Learning (TML)	The ability of a computer to learn from data streams by using automated machine learning applied to the entire, or partial, data stream set that leads to a frictionless machine learning process that is continuous and mostly uninterrupted by humans.
Training dataset	A consolidation of data streams made up of a dependent variable stream and independent variable stream(s) that are in a format that allows a computer to apply algorithms to find and learn patterns and underlying structures and relationship in the data.

References

1. Aggarwal, C.C. (2007). Data Streams: Models and Algorithms, Springer.

2. Agrawal, R., T. Imielinski, A. Swami (1993), "Mining Association Rules between Sets of items in Large Databases," ACM SIGMOD Conference.

3. Amershi, S., M. Cakmak, W.B. Knox, T. Kulesza (2014), "Power to the people: The role of humans in interactive machine learning," AI Magazine.

4. Bolton, R. and D. Hand (1999), "Unsupervised profiling methods for fraud detection," In Credit Scoring and Credit Control VII.

5. Cormode, G., and S. Muthukrishnan (2005), "An improved data stream summary: the count-min sketch and its applications," Journal of Algorithms, Volume 55, Issue 1, Pages 58–75.

6. Datar, M., A. Gionis, P. Indyk, R. Motwani (2002), "Maintaining stream statistics over sliding windows," SIAM Journal on Computing 31(6): 1794–1813.

7. Davenport, T. (2006), "Competing on Analytics," Harvard business review 84(1): 98–107.

8. Duan, Y., John S. Edwards, Yogesh K Dwivedi (2019), "Artificial intelligence for decision making in the era of Big Data – evolution, challenges and research agenda," International Journal of Information Management, Volume 48, Pages 63–71.

9. Eugster, P. TH., Pascal A. Felber, Rachid Guerraoui, Anne-Marie Kermarrec (2003), "The Many Faces of Publish/Subscribe," ACM Computing Surveys, Vol. 35, No. 2, June 2003, pp. 114–131.

© Sebastian Maurice 2021
S. Maurice, *Transactional Machine Learning with Data Streams and AutoML*,
https://doi.org/10.1007/978-1-4842-7023-3

10. Faiz, R.B., Eran A. Edirisinghe (2009), "Decision Making for Predictive Maintenance," Asset Information Management Interdisciplinary Journal of Information, Knowledge, and Management Volume 4, 2009.

11. Fei, Xiang, Nazaraf Shah, Nandor Verba, Kuo-Ming Chao, Victor Sanchez-Anguix, Jacek Lewandowski, Anne James, Zahid Usman (2019), "CPS data streams analytics based on machine learning for Cloud and Fog Computing: A survey," Future Generation Computer Systems, Volume 90, 2019, Pages 435–450, ISSN 0167-739X, https://doi.org/10.1016/j.future.2018.06.042

12. Garofalakis, Minos and Gehrke, Johannes. (2002). "Querying and Mining Data Streams: You Only Get One Look," Conference: VLDB 2002, Proceedings of 28th International Conference on Very Large Data Bases, August 20–23, 2002, Hong Kong, China.

13. Gudivada, V., A. Apon, and J. Ding. "Data Quality Considerations for Big Data and Machine Learning: Going Beyond Data Cleaning and Transformations." In: International Journal on Advances in Software 10.1 (2017), pp. 1–20.

14. Guzy, F., M. Woźniak (2020), "Employing dropout regularization to classify recurring drifted data streams," 2020 International Joint Conference on Neural Networks (IJCNN), Glasgow, United Kingdom, 2020, pp. 1–7, doi: 10.1109/IJCNN48605.2020.9207266.

15. Hand, David, Heikki Mannila, and Padhraic Smyth (2001), Principles of Data Mining. MIT Press, Cambridge, Massachusetts. ISN 0-262-08290-X.

16. Hartman, R.S. (1969), The Structure of Value (Carbondale, Southern Illinois University Press, 1969 (paperback edition), p. 154.

17. Hartman, R.S. (1972), "The Value Structure of Creativity," The Journal of Value Inquiry, Vol. VI, No. 4, 1972, p. 250.

18. He, Xin, Kaiyong Zhao, Xiaowen Chu (2020), "AutoML: A Survey of the State-of-the-Art," `https://arxiv.org/abs/1908.00709v5`

19. Heravizadeh, M., J. Mendling, et al. (2009). Dimensions of business processes quality (QoBP), Springer.

20. Jayanthi, M.D., G. Sumathi (2016), "A Framework for Real-time Streaming Analytics using Machine Learning Approach," Proceedings of National Conference on Communication and Informatics-2016, Organized by Department of Information Technology, Sri Venkateswara College of Engineering, Sriperumbudur.

21. Jayanthiladevi, A., Surendararavindhan, Sakthivel (2018), "Fast data vs. big data with iot streaming analytics and the future applications," in Handbook of Research on Cloud and Fog Computing Infrastructures for Data Science: IGI Global, 10.4018/978-1-5225-5972-6.ch016.

22. Kaplan, R.S., D.P. Norton (1996), The balanced scorecard. Translating strategy into action. Harvard Business School Press, Boston.

23. Kaplan, R.S., D.P. Norton (2001), The strategy-focused organization. How balanced scorecard companies thrive in the new business environment. Harvard Business School Press, Boston.

24. Kerr N.L., R.S. Tindale (2004),"Group performance and decision making," Annu Rev Psychol, 55:623–55.

25. Kollios, G., Papadopoulos, D., Gunopulos, D. (2005) "Indexing mobile objects using dual transformations," The VLDB Journal 14, 238–256.

26. Kugler T, E.E. Kausel, M.G. Kocher (2012), "Are groups more rational than individuals? A review of interactive decision making in groups," Wiley Interdiscip Rev Cogn Sci., 3:471–482.

27. Lohr, Steve (2013), "The Origins of 'Big Data': An Etymological Detective Story." The New York Times.

28. Looy, A.V., Aygun Shafagatova, "Business process performance measurement: a structured literature review of indicators, measures and metrics," SpringerPlus (2016) 5:1797.

29. Lozano, A.M., Nir Lipsman, Hagai Bergman, Peter Brown, Stephan Chabardes, Jin Woo Chang, Keith Matthews, Cameron C. McIntyre, Thomas E. Schlaepfer, Michael Schulder, Yasin Temel, Jens Volkmann, Joachim K. Krauss (2019), "Deep brain stimulation: current challenges and future directions," Nat Rev Neurol, 15(3): 148–160.

30. Mashey, John R. (1999), "Big Data ... and the Next Wave of InfraStress," www.usenix.org/conference/1999-usenix-annual-technical-conference/big-data-and-next-wave-infrastress-problems

31. Maurice S., Ruhe G., Saliu O., Ngo-The A. (2006), "Decision Support for Value-Based Software Release Planning," In: Biffl S., Aurum A., Boehm B., Erdogmus H., Grünbacher P. (eds) Value-Based Software Engineering. Springer, Berlin, Heidelberg.

32. McCormick, J. (2020), "AI Project Failure Rates Near 50%, But It Doesn't Have to Be That Way, Say Experts," Wall Street Journal, Aug. 7, 2020.

33. McGilvray, D.. (2008). Executing Data Quality Projects.

34. Mitchell, T. (1997), Machine Learning, Springer, 1997.

35. Mohammadi, M., A. Al-Fuqaha, S. Sorour and M. Guizani (2018), "Deep Learning for IoT Big Data and Streaming Analytics: A Survey," in IEEE Communications Surveys & Tutorials, vol. 20, no. 4, pp. 2923–2960.

36. Pisano, G. (2015), "You need an innovation strategy," Harvard Business Review.

37. Read, J., A. Bifet, W Fan (2019), "Introduction to the special issue on Big Data, IoT Streams and Heterogeneous Source Mining." Int J Data Sci Anal 8, 221–222 (2019).

38. Read J., R.A. Rios, T. Nogueira, R.F. de Mello (2020), "Data Streams Are Time Series: Challenging Assumptions," In: Cerri R., Prati R.C. (eds) Intelligent Systems. BRACIS 2020. Lecture Notes in Computer Science, vol 12320. Springer, Cham. https://doi.org/10.1007/978-3-030-61380-8_36

39. Sagiroglu, S. and Sinanc, D. (2013) Big Data: A Review. 2013 International Conference on Collaboration Technologies and Systems (CTS), San Diego, 20–24 May 2013, 42–47.

40. Sakurai, Y., S. Papadimitriou, D. Faloussos (2005), "BRAID: Stream mining through group lag correlations," ACM SIGMOD Conference.

41. Samuel, Arthur L. (1959). "Some Studies in Machine Learning Using the Game of Checkers." IBM Journal of Research and Development. 44: 206–226. Carter, B. (2013).

42. Saaty, Thomas L. (1982), Decision Making for Leaders: The Analytical Hierarchy Process for Decisions in a Complex World, Belmont, California: Wadsworth. ISBN 0-534-97959-9; Paperback, Pittsburgh: RWS. ISBN 0-9620317-0-4.

43. Saaty, Thomas L., Kirti Peniwati (2008), Group Decision Making: Drawing out and Reconciling Differences. Pittsburgh, Pennsylvania: RWS Publications. ISBN 978-1-888603-08-8.

44. Shaw M.E. (1932), "A comparison of individuals and small groups in the rational solution of complex problems," Am J Psychol, 44:491–504.

45. Sidi, F., P. H. Shariat Panahy, L. S. Affendey, M. A. Jabar, H. Ibrahim and A. Mustapha (2012), "Data quality: A survey of data quality dimensions," 2012 International Conference on Information Retrieval Knowledge Management (CAMP), pp. 300–304, 2012.

46. Soni, Neha, Enakshi Khular Sharma, Narotam Singh, Amita Kapoor (2020), "Artificial Intelligence in Business: From Research and Innovation to Market Deployment," Procedia Computer Science, Volume 167, pp. 2200–2210.

47. Waller, D. (2020), "10 Steps to Creating a Data-Driven Culture," Harvard Business Review, Feb. 2020.

48. Weiss, Sholom M. and Nitin Indurkhya (1998), Predictive data mining: a practical guide. Morgan Kaufmann Publishers Inc. San Francisco, CA, United States.

49. Williams, M.R. (2000), "A Preview of Things to Come: Some Remarks on the First Generation of Computers," in *The First Computers—History and Architectures*, edited by Raúl Rojas and Ulf Hashagen, 2000: MIT Press, Cambridge, Mass.

50. Wrench, C., F. Stahl, G. Di Fatta, V. Karthikeyan, D.D. Nauck (2016), "Data stream mining of event and complex event streams: a survey of existing and future technologies and applications in big data," In: Atzmueller, M., Oussena, S. and Roth-Berghofer, T. (eds.) Enterprise Big Data Engineering, Analytics, and Management. IGI Global, pp. 24–47 ISBN 9781522502937 doi: https://doi.org/10.4018/978¬1¬5225¬0293¬7 Available at http://centaur.reading.ac.uk/68050/

51. Yao, Q., M. Wang, Y. Chen, W. Dai, H. Yi-Qi, L. Yu-Feng, T. Wei-Wei, Y. Qiang, Y. Yang (2019), "Taking human out of learning applications: A survey on automated machine learning," arXiv:1810.13306.

52. Yi, B.-K., N.D. Sidiropoulos, T. Johnson, H.V. Jagadish, C. Faloutsos, A. Biliris (2000), "Online data mining for co-evolving time sequences," ICDE Conference.

53. Zolas, Nikolas, Zachary Kroff, Erik Brynjolfsson, Kristina McElheran, David Beede, Catherine Buffington, Nathan Goldschlag, Lucia Foster, Emin Dinlersoz (2020), "Advanced Technologies Adoption and Use by U.S. Firms: Evidence from the Annual Business Survey," US Census Bureau.

Index

A

W, X, Y, Z

Printed in the United States
by Baker & Taylor Publisher Services